THE COAL VIEWER

AND

ENGINE BUILDER'S PRACTICAL COMPANION

THE COAL VIEWER

AND

ENGINE BUILDER'S PRACTICAL COMPANION

BY

JOHN CURR

With a new introduction by Charles Lee

SECOND EDITION

Reprints of Economic Classics

AUGUSTUS M. KELLEY PUBLISHERS

New York, 1970

Published by
FRANK CASS AND COMPANY LIMITED
67 Great Russell Street, London WC1

Published in the United States by
Augustus M. Kelley, Publishers
New York, New York 10010

First edition	1797
Second edition	1970

ISBN 678 05104 6

Library of Congress Catalog Card No. 74—96376

Printed in Great Britain

INTRODUCTION TO THE SECOND EDITION

THE one common characteristic of the many forms of track which may be regarded as a railway is that the combination of track and vehicle is self-steering. A projection from the vehicle to engage the inner faces of the rails is the earliest known form, and is clearly illustrated by Georgius Agricola in *De Re Metallica* in 1556. This was succeeded by the flange on the wheels, as in common railway practice, but this had (and has) the limitation that the vehicles are unfitted for running on the highway. The great merit of the work in this sphere of John Curr, the author of *The Coal Viewer, and Engine Builder's Practical Companion*, is that he devised for the coal mining industry a form of "cast iron rail road" to accommodate flat-tyred vehicles capable of being used apart from the railway. Moreover, the section of his book devoted to this device is the earliest known engineering work on railway track.

During the material part of his career, John Curr worked in Sheffield, but he is generally believed to be a native of the County of Durham, the son of John Curr, coal viewer, Bushblades, who was buried at Lanchester on 24 April 1777. The actual birthplace of our John Curr has not been determined. In *A History of the Parish of Dipton* (Durham, 1911), J. W. Fawcett summarised the information so far as it was revealed by his own local knowledge and extensive research, when he stated "One account says he [John Curr] was born at Kyo, near Lanchester, and brought up at Pontop Pike; and another that he was born and brought up at Greenside in Ryton Parish, but very little is definitely known. . . . He was born in the year 1756, but there is no entry of his baptism in the Parish Registers at Lanchester, so can scarcely be said to have been born in that parish, and Kyo was part of Lanchester parish at the date of his birth." The negative evidence is not so conclusive as Fawcett seems to have thought, as we have no knowledge of the religious allegiances of the Curr family at the time; later, our author was a prominent Roman Catholic in Sheffield and was buried at a Roman Catholic Church there, St. Marie's, Norfolk Row.

In 1934, Grafton & Company's catalogue described *The Coal Viewer* as a "very rare and important work," and certainly the number of known copies is small. They all bear the date 1797, but the late Mr. C. F. Dendy Marshall once possessed a copy (possibly a proof) dated 1796. In his 'introductory preface,' Curr states that "the making and use of rail roads and corves were the first of my inventions, and were introduced at the Sheffield Colliery about twenty-one years ago." This means that he was working in Sheffield about 1776 when still a young man, and the introduction of his device was a remarkable achievement, even allowing for the fact that there is no reason to assume that it was at first developed to anything approaching the perfection which he described in this book two decades later. There is a cryptic reference to the invention in a book published by his son, John Curr Junior, in 1847, which says of the inventor "his father differed with him in opinion as to the result of the invention, and disowned him, and the letter to that effect was retained in his family for 30 years or more."

The Sheffield Colliery was owned by the Duke of Norfolk (to whom Curr dedicated *The Coal Viewer*), but was leased until 1780 to Townsend and Furniss, and it was during this lease that a waggonway or "Newcastle road" was laid between the colliery and Sheffield, in 1774. As the words convey, this was a line with wooden rails on wooden transverse sleepers, to accommodate flanged-wheeled wagons, and there is no evidence to suggest that John Curr was associated with its construction or indeed had arrived in Sheffield at the time, despite numerous allegations to this effect by modern writers. From 1780, the Duke of Norfolk worked his own coal, with Curr as the "Superintendent of the Coal Works." Initially, Curr's rails were used only in underground workings, and tribute to their efficiency was paid by John Buddle Senior (1743–1806) in his "Report on Sheffield Park and Attercliffe Collieries" dated 7 April 1787. Curr himself stated that 12 years passed before his rail roads were much imitated, but then their use spread extensively. In *A General View of the Agriculture of Derbyshire* (1817), John Farey Senior (1766–1826) stated "I have hear it said, that the earliest use of these flanched Rails above ground (for they were first introduced in the underground Gates of Mines, it is said) was on the South of Wingerworth Furnace; leading to the Ironstone Pits, by Mr. Joseph Butler, about the year 1788." The century and a half which have elapsed since Farey was writing have failed to produce firm evidence of an earlier date than 1788 for the first flanged plate rail above ground.

Because of the poor financial results achieved by the Duke of Norfolk's collieries towards the end of the eighteenth century, Curr was suddenly dismissed in October 1801, without any reason being given to him. A long letter of protest which he wrote to the Duke on 23 October 1801 recited the numerous inventions and improvements which he had introduced. With regard to his rail roads, he said "about 16 collieries out of 20 have introduced this mode of conveying coals in the Countys of York, Lancaster, Salop, Derby, Stafford, Warwick, and great part of Wales, and is now adopting near London and in the neighbourhood of Newcastle upon Tyne." This is an important statement, and obviously refers to many lines laid by Benjamin Outram (1764–1805), who often has been credited wrongly with the invention of flanged plate rails. Although he used them extensively, he himself made no such claim, and an interesting and little-known manuscript, undated, but probably about 1803 or 1804, preserved in the Outram family is "Minutes to be Observed in the Construction of Railways." In this he mentions the flanged-plate lines "on a system introduced by Mr. Curr." In the *Transactions of the Highland Society* (1824), an important and informative Paper by Alexander Scott of Ormiston says that, for "the introduction of cast-iron railroads below ground . . . the mining industry is indebted to Mr. John Carr [Curr], who introduced them, about the year 1776. into his Grace the Duke of Norfolk's works, near Sheffield."

Curr's ingenuity was not confined to his plate rails. He invented conductors or guides, or vertical railways in shafts, for which he secured Letters Patent No. 1660, dated 12 August 1788; invented flat hemp ropes covered by Letters Patent No. 2270 dated 17 November 1798; and was a pioneer in employing fixed steam engines for hauling wagons, about 1805. With the approval of the Duke of Norfolk, he had established a foundry in 1792, and an Account Book of 1805 shows that the Duke of Norfolk continued to buy castings and flat ropes from him. About 1785 Curr married Hannah Wilson (a native of Sheffield, who died at York on 10 June 1851) and had two sons and five daughters. It appears that Curr and his family went to France for a time in 1820, but he returned to Sheffield and died there at Belle Vue on 27 January 1823.

One of the sons (also John) went to New South Wales and in September 1847 published a book entitled *Railway Locomotion and Steam Navigation: Their Principles and Practice*. This book, already mentioned briefly, referred to *The Coal Viewer*, and said of his father "He was the inventor of the cast-iron railway, then called a railroad; in the book are plans of every description of rail, with particulars of their length on passing curves, given to the one hundreth part of an inch. He therein foretels (*sic*) what in time it would become; but note, the plan was opposed; it meddled with the interest of a certain body,—a riot ensued,—the railway was torn up, the coal staith burnt, and the inventor, my father, reduced to the necessity of concealment in a wood, for three days and nights, to escape the fury of the populace."

The remainder of *The Coal Viewer* is taken up with the most elaborate description of a Newcomen pumping engine, at the highest state of its development. The tabular material is extraordinarily detailed; few modern writers would regard it as necessary to carry to the fourth decimal place the quantity of water in pumps. The present-day reader should not be deterred by the use of the long "s" as it is a noteworthy book in numerous respects, written by a man who was well to the fore in practical ingenuity. It is to be regretted that a full-scale account of his activities is not available. A Paper on John Curr by Councillor Fred Bland of Sheffield was presented to the Newcomen Society in June 1931, but this is now in need of amplification (and, in some respects, amendment) in the light of more recent research, such as that revealed by Dr. R. A. Mott in an unpublished lecture which he delivered to the Sheffield Trades Historical Society on 7 March 1966, and developed in a comprehensive survey of John Curr's work in the sphere of mining transport presented to the Newcomen Society on 8 October 1969.

CHARLES E. LEE

November 1969

THE

COAL VIEWER,

AND

ENGINE BUILDER's

PRACTICAL

COMPANION.

BY JOHN CURR, OF SHEFFIELD.

SHEFFIELD:

PRINTED FOR THE Author, BY JOHN NORTHALL.

And Sold by

I. & J. TAYLOR, *at the* ARCHITECTURAL LIBRARY, HIGH HOLBORN, *London*; JOSEPH WHITFIELD, *Newcaſtle-upon-Tyne*; ARCHIBALD CONSTABLE, *Edinburgh*; DUNLOP & WILSON, *Glaſgow*; I. & W. EDDOWES, *Shrewſbury*; AND JAMES HARROP, *Mancheſter*.

ENTERED *at* STATIONER's HALL.

1797.

TO

THE MOST NOBLE PRINCE,

CHARLES HOWARD, DUKE OF NORFOLK,

Earl Marſhal

AND HEREDITARY EARL MARSHAL

Of ENGLAND;

EARL OF ARUNDEL, SURREY, NORFOLK AND NORWICH,
&c. &c. &c.

THIS

COAL VIEWER,

AND

ENGINE BUILDER's PRACTICAL COMPANION,

IS, (WITH PERMISSION)

MOST HUMBLY DEDICATED,

By His Grace's

Moſt Obedient

Humble Servant,

JOHN CURR.

INTRODUCTORY PREFACE.

SHOULD the following ſheets meet with approbation from the public, equal to the ſucceſs, which has attended the execution of the various articles deſcribed in their contents, the object of my wiſhes will be fully attained;—and, as no work of the like kind has hitherto been publiſhed, I flatter myſelf this will be found uſeful.

It ſeldom falls to the lot of literary men, to be engaged in works of this ſort, and therefore profeſſing myſelf to be merely a mechanic, it can ſcarcely be expected that I ſhould convey my ideas in all the elegance of expreſſion, of which our language is capable; beſides, it muſt occur to every reader, that ſuch a work as this will not admit of any great choice of words, when it is conſidered, that by far the greater part conſiſts of the various ſynonymous technical terms uſed in different parts of the kingdom, and of which, in order to convey a clear idea to every claſs of readers, there are unavoidably frequent repetitions and explanations, and if I have the happineſs to make myſelf underſtood on this head, I hope it is all that will be required of me.

The tables and eſtimates of this work are the reſult of upwards of twenty years ſtudy and practice, and are extracted from calculations and obſervations, which would have filled ſeveral volumes.——They were originally compoſed for my private uſe and that of my aſſiſtants, in order to facilitate and diſpatch buſineſs.—The greateſt part of them have been uſed ſeveral years, and where any error has been diſcovered it has been rectified, ſo that I have the ſatisfaction to aſſure the public their accuracy may be relied on.

The making and uſe of rail-roads and corves were the firſt of my inventions, and were introduced at the Sheffield Colliery about twenty-one years

years ago; they are doubtlefs a great acquifition in rendering the article of conveyance much eafier and lefs expenfive, and it is not the leaft convincing proof of their being fo, that they have been generally imitated, and made ufe of in moft collieries for the laft three years, efpecially in the fouthern parts of the kingdom.

The table fhewing the quantity of coals contained in an acre, is full as accurate as the fubject requires, and the weight (which varies a little) will be found to anfwer in the average of coal throughout the kingdom.

The various names which I have introduced to diftinguifh the articles that compofe the Steam Engine, may not be intelligible to every one, but as I have in general given a reference to the plate which fhews the form of the figure alluded to, the difficulty will be explained; and for the greater difpatch, I have alfo introduced a reference in the Steam Engine tables to the page which explains them, as alfo a reference from the page that explains them, to the page of the table where the fundry fizes and dimenfions of the articles fuitable to all Engines are to be found.

THE

COAL VIEWER,

&c.

ON

CONVEYING COALS UNDERGROUND.

THE Collieries which were opened throughout this kingdom in preceding ages, being in a great meaſure exhauſted in the *baſſet*, *crop*, or *outbreak* coal, and ſuch coal as lay within moderate depths of the ſurface, it has become neceſſary to eſtabliſh works at a greater depth, and in conſequence to ſink the pits at greater diſtances from each other, (which without improvements are made,) muſt increaſe the expence of conveying the coals to the bottom of the pits; and the drawing of coals up the ſhafts by machines being rendered practicable a few years ago, *which* affords a very conſiderable ſaving in that article, where the depths are great: *this* muſt of courſe, (to avoid the great expence of ſinking pits and removing the machines,) point out the neceſſity of conveying the coals underground from

from a greater diſtance than uſual; hence it becomes a meaſure of great importance in collieries, to contrive the moſt *eaſy* and *expeditious* mode of ſo conveying the coals *underground.*

The prevailing practice, till of late, in the working of collieries in the neighbourhood of Newcaſtle-upon-Tyne and Sunderland, was to draw a ſingle corf* on a ſled from the workings to the ſhaft of the pit, which as theſe workings were extended, and the prices and maintenance of horſes enormouſly encreaſed, became an intolerable burthen to the proprietors of ſuch works; therefore the viewers or ſuperintendents of collieries, have with a great deal of propriety introduced wooden rails, or waggon ways underground, for that purpoſe, (or what is generally diſtinguiſhed by the name of Newcaſtle-roads,) and fixed a frame upon wheels capable of receiving *two or three* of their *baſket corves*, which upon theſe carriages and roads are drawn by *one* horſe. But the baſket or twig corf which has ſome great perfections to recommend it at Newcaſtle-upon-Tyne and Sunderland, where the coals are ſmall, (being of a globular form, with a ſmall aperture at the top,) cannot with any propriety be introduced in the ſouthern parts of this kingdom, where the coals delivered to market are all, or in a great meaſure, large. And

* A machine made of wood or twigs, in which the coals are drawn from the face of the vein or bed, to the bottom of, and up the ſhafts.

And notwithſtanding this great improvement, I am of opinion that a greater acquiſition is ſtill to be made with the ſame corf, by laying *caſt iron roads* upon the plans hereafter ſet forth, and placing the corf upon a ſmall *frame* or *tram* made upon a proper principle, and hooking or chaining one tram to another, as a view of them plate the 3d, fig. the 3d, 4th, and 8th points out, which ſhews the wheels both in the inſide and outſide of the frame.

Having for the above mentioned reaſons introduced machines for *drawing* coals at two of His Grace the Duke of Norfolks Collieries, near Sheffield, I had ſtill a difficult point to accompliſh, *which was,* to contrive an *eaſy* and *expeditious* mode of *conveying* the coals to the bottom of the pit, in which I have been ſucceſsful, far beyond my expectations, and perhaps have hit upon a mode ſuperior to any thing heretofore practiſed, as the reſult of ſeven years experience informs me; I have therefore herein offered to the public the plans and directions for executing both the roads and corves, and every thing relating to the invention, by which means a horſe takes at a moderate draught, nine or ten corves of equal ſize to thoſe at Newcaſtle-upon-Tyne and Sunderland, of which, even by their improved mode of *conveying*, the horſe takes only two or three.

This is however not the only advantage attending theſe

caſt iron roads and trams; for by the mode of *conveying* at Newcaſtle-upon-Tyne and Sunderland, the weight of the two or three loaded corves, together with that of the frame on which they are placed, is ſupported upon a very ſhort ſpace of the road, (perhaps four or five feet only,) and by my mode of doing it the ſame quantity of coals is diſperſed upon ſix times that ſpace; which muſt of courſe take off all unreaſonable preſſure, and render the roads more durable. There are in my opinion many ſituations in this kingdom, not only in the coal trade and underground, but for the *conveying* of goods in fixed ſituations of any kind above-ground, which *can* conveniently be reduced into parcels or portions of five or ten hundred weight each, (or even twenty or thirty if the roads are made a little ſtronger,) and where the ſurface of the ground for conveyance lies tolerably level, or of an eaſy deſcent, that the caſt iron roads and carriages upon the principle of thoſe hereafter deſcribed, may with great advantage be introduced. And indeed where a mode of conveyance is wanted upon *a declining* plane of any deſcent exceeding three inches to the yard, theſe roads and carriages far exceed any other mode. As all the coals underground to the *riſe* or *baſſet* of the before mentioned collieries are conveyed upon this principle, where the weight and velocity of the loaded carriage acquired in going down the *declining* plane, take up the empty one, experience

experience enables me to ſpeak to this with preciſion; and the curious may, there, have occular proof of the ſame.

I have adopted this mode of *conveying* coals *above* the ground alſo, for *ſtacking* them, and find, the ſcheme is manageable for three hundred yards diſtance.

The expence of ſupporting roads and carriages upon theſe principles, I am perfectly convinced from experience, is trifling, in compariſon with either the turnpike road, or wooden rail or waggon ways; the execution effected, and the eaſe of conveyance are certainly proved to be ſuperior, and the firſt expence is demonſtrably leſs than by either the double rail waggon way or turnpike road; and in regard to collieries in the ſouthern parts of this kingdom, where the *breakage* of *coals* is of conſequence, the *loſs* ſuſtained in conveying in large quantities, (ſuch as waggon loads,) and in the unloading and loading again into other carriages, is alone a ſufficient recommendation.

Some of my readers will perhaps ſtart an objection and ſay, how would you go round the ſquare turns underground with ſuch long draughts of corves as theſe? To which I anſwer, that in the main roads underground, where the *hurrying*, *putting* or *conveying* by the *horſe* is introduced, ſquare turns are not neceſſary; that where turns

are

are needful, by taking room for a regular curve line, they are rendered very practicable, as the many turns that inevitably attend our preſent underground roads ſufficiently demonſtrate. The *weight* and *expence* of the *corf* will perhaps alſo be objected to; in regard to *weight*, in the time of *drawing*, one *corf* will always counterbalance another; and in point of *uſe*, I muſt take the liberty of obſerving, that the modes I have invented of ſtriking, or landing and emptying them, (for which, along with ſome other improvements I have obtained his Majeſty's Royal Letters Patent,) are rendered perhaps more manageable than any other corf, (the baſket corf excepted;) and with reſpect to *expence*, being ſo perfectly preſerved by my invention of *conductors*, during their being drawn up a ſhaft, by the extreme velocity of machines, (even through a ſpace of one hundred and forty yards, in half a minute,) and the margin of the caſt iron roads underground preſerving them always from wearing againſt the ſides of the gates or ways, their neceſſary repairs are ſo trifling, and their duration ſo great, as almoſt to ſurpaſs conception or belief; the greateſt part of the corves at theſe collieries has been in common uſe for five or ſix years, and when examined will be found little *worſe for the wear*.

Theſe roads and corves are alſo applied in ſundry collieries where barrowmen only are introduced; and might in in my opinion be extremely ſerviceable to ſundry large lime works in Staffordſhire and Shropſhire. At Froghill, in Staffordſhire, they have a land conveyance for their limeſtone, which is three or four miles in length, one half of which is a flat ground, and the other half, about two and a half, or three inches deſcent in the yard; theſe roads, which are upon the plan of what is called Newcaſtle waggon roads, are laid in a firm manner upon wood, (after having been at a great expence of ſtoneing about ten or twelve inches thick for a foundation;) upon this wood is laid caſt iron an inch and a half thick, a part of which weighs in every ſingle yard forward one hundred and forty-one pounds, and other models weigh only eighty-one pounds: when the waggons come upon theſe roads, which together with the limeſtone weighs in the ſundry kinds of theſe carriages, they do, and have made uſe of, not leſs than four, five, and ſix tons, and I believe as much as ſeven tons even, which burden being laid all upon four feet in length, the above roads, although enormous in the firſt expence, are nothing too ſtrong. Were my roads and carriages introduced in ſituations ſimilar to this, where there is nothing wanted in the road but caſt iron plates half an inch thick, (one yard forward of which road weighs about forty-eight pounds,) and a ſleeper of wood, four inches by two and a half, at every two yards aſunder,

 and

and a ſmall carriage upon the conſtruction of our corves, by which means the draught of the horſe would be diſperſed upon twenty yards, inſtead of four feet, the ſavings would be very conſiderable indeed; not to mention, that inſtead of applying a friction upon the waggon wheel to hold them down the hills, and dragging the empty ones back again by horſes, they might take the opportunity of making the full carriage down hill, take back the empty one, upon the ſame principle as we convey our coals down the gates or ways underground at Sheffield and Attercliffe collieries.

That my readers, who are acquainted with the conveying of coals underground, may be enabled to compare my mode of conveyance with what they have ſeen practiſed, I muſt inform them, that our corves are collected together on the ſides of the main road, (which is nearly a dead level,) in four or five different parts of the works, until they amount to the number of 11, 12, 13, or 14 in a place; that each corf contains nineteen pecks Newcaſtle meaſure, in weight about $5\frac{1}{2}$ cwt. and that a horſe conveys for a moderate days work, the quantity of one hundred and fifty tons, the diſtance of two hundred and twenty yards, by taking in general twelve of theſe corves at a draught, but where the ground deſcends half an inch in the yard, a horſe will take double that quantity.

THE CORF.

THE Corf fully disfected, plate 1ft, figures 1, 2, 3, 4, 5, 6, 7, 8, 9, 10, 11, 12, 13 and 14, is adapted chiefly to the long way of working collieries, where the roads along the *benk* faces are narrow, and where the coals are chiefly fmall, weighs itfelf about twenty-four ftones, and when filled up with large coals, carries about nineteen or twenty pecks Newcaftle meafure, in weight 5½ or 6 cwt. Fig. 1, fhews the end, fig. 2, the fide, and fig. 3, the bottom of the corf when turned upwards.

DIMENSIONS.

Length on the outfide 40 inches, breadth 30; infide length 38 inches, breadth 27½ inches, and height 21½ inches, and ftands 30 inches high upon the wheels; contains neat meafure clear of the boxes which cover the wheels, to the ftreak 20491 folid inches.

DIRECTIONS to the SAWERS.

The fides and ends to be Afh; the fides fhould be fet out 41½ or 42 inches long, and repeated as often as fuits the timber; cut 1½ inch thick at one edge, and 1 inch at the other, clear from the faw, and 26 inches deep, which depth

depth may be made up by two pieces, and *dowelled* together by the carpenter, if one piece is not found deep enough. The ends ſhould be cut $31\frac{1}{2}$ inches long, 21 inches deep, and 1 inch thick, clear from the ſaw; boxes to cover the wheels, of deal board $\frac{3}{4}$ inch thick; flags for the corf bottom, of Oak 39 inches long; bars or ſpendings of Oak, cut $31\frac{1}{2}$ inches long, and riven $3\frac{3}{4}$ inches broad and 2 inches thick.

DIRECTIONS to the CARPENTER.

Be very careful to cut out both ends and ſides by models, and frame the corf exactly to 40 inches long and 30 inches broad, (outſide dimenſion,) or the rolled iron will not fit. The ends muſt be counterſunk $\frac{1}{4}$ of an inch deep, and the ſides cut to joint into the ends, as the bottom of the corf, plate 1ſt, and fig. 3, deſcribes.

The caſt iron buſhes ſhewn, front and ſide view, plate 1ſt, and fig. 4, the holes of which are full $1\frac{1}{4}$ inch, inſide diameter, are let into the ſides, and riveted to it, before the corf is put together, and are fixed $16\frac{5}{8}$ aſunder, from center to center, and the center of them is placed two inches from the edge of the wood.

The top ſide of the ſpending holes, which are 2 inches by 1 inch, are ſet out $3\frac{3}{4}$ inches from the edge of the wood, and

and placed juſt ſo far aſunder as to leave 2 inches of ſolid wood from the end, which ſituation leaves only ſufficient room for the wheels to work clear, and the ſpendings are ſhouldered on one ſide.

DIRECTIONS to the BLACKSMITH.

The blackſmith muſt be very careful to make his axletrees plate 1, fig. 5, to the exact length of 2 feet $7\frac{1}{2}$ inches, and $1\frac{1}{8}$ diameter when turned in the lathe; the part which muſt be turned, muſt meaſure to the ſhoulder of the boſs $4\frac{1}{2}$ inches, the boſs muſt be $\frac{1}{2}$ an inch broad, thence to the edge of the cotter hole which is $\frac{5}{8}$ inch broad, and $\frac{1}{8}$ thick, muſt be 1 foot 9 inches, and from the cotter hole to the end muſt be $4\frac{7}{8}$ inches.

The hoop (plate 1, fig. 7, gives an end and ſide view,) which forms an inſide ſhoulder for the wheel, muſt be malleable iron, $\frac{1}{4}$ of an inch thick, and $1\frac{3}{4}$ long, with a cotter hole through it $\frac{5}{8}$ of an inch by $\frac{1}{8}$, and this end of the axletree muſt be turned 9 inches in length, that the hoop may ſlide along it, for the convenience of putting on the wheels.

Should the corves be made to draw by conductors, the chains by which they are ſuſpended muſt be made of an exact length; and the links ſhould be $1\frac{5}{8}$ of an inch circumference; and from the center of the tug hole to the center of the ring that connects them, ſhould meaſure 22½ inches: and if the corves be intended to be put or hurried by horſes, the links for connecting them muſt be made 1½ inch circumference, and the hook in the turn to be made flat, and very ſtrong; the three links and hook muſt weigh 3lb. if they be ſtrong enough, and ſhould meaſure 11½ or 12 inches extreme length.

The rolled irons for the corf require exactneſs in the firſt ſetting out, but after a ſingle model of each is made, it will be found very convenient for marking the reſt by.

Fig. 8, is one of the angular bottom plates, the whole length of which is 4 feet 8¾ inches, breadth 2 inches, and bare ⅛ thick; the firſt hole is 1¼ inch from the end, and the middle hole will be found 2 feet 3⅛ inches.—Fig. 9, is the plate which lies along the bottom, the whole length 3 feet 7 inches, is full ⅛ of an inch thick, and 2¾ broad; the end holes are 1¼ inch from the end, (meaning in this and all other caſes where I mention holes, to give the diſtance to the centers of them,) and the next hole 4 inches farther, and to the middle hole it will be found to be 16¼ inches.—Fig. 10, is the plate that goes up each corner of the corf; is 4 feet long, ⅞ of an inch broad, and 5-16 thick, and meaſures 2 feet 2½ inches from the turn to the center

center of the hole.—Fig. 11, meaſures 3 feet 10$\frac{5}{8}$ inches long, 2 inches broad, and bare $\frac{1}{8}$ thick, and is the plate which croſſes over the end of the corf; the firſt hole is 1$\frac{1}{4}$ inch from the end, and will be found 3 feet 8$\frac{1}{8}$ inches aſunder, and the ends are bent down 2$\frac{1}{2}$ inches from the ſtraight line, at 6$\frac{1}{2}$ inches from the end.—Fig. 12, is the corner plate, the whole length of which is 15$\frac{1}{2}$ inches, and 2 inches broad, and bare $\frac{1}{8}$ thick; the holes are 9$\frac{1}{2}$ inches aſunder, and the turn of the plate falls in the middle of it.—Fig. 13, a ſhort plate 9 inches long, 2 inches broad, and bare $\frac{1}{8}$ thick, and the hole is placed 2 inches from the end: this plate is nailed under the ſpending of the corf, and is brought through the mortiſe in the ſide. All the ſquare holes in this rolled iron are $\frac{1}{2}$ an inch ſquare, and the round ones are $\frac{1}{2}$ an inch diameter.—When the rolled irons are placed upon the corf in their ſeveral ſituations, the holes are filled up with $\frac{1}{2}$ inch ſquare or round bolts. Every other article of blackſmiths work in the corf will readily be diſcovered by inſpecting the drawing. If the wheel will juſt turn round upon the axletree it is quite ſufficient, but the buſhes ſhould have a full $\frac{1}{8}$ inch play.—Fig. 14, is the rolled iron by which the corf is ſuſpended; it is 4 feet long, 2 inches broad, and bare $\frac{1}{8}$ thick; the firſt hole is 1$\frac{1}{4}$ inch from the end, the ſecond meaſures 14$\frac{1}{4}$ inches, and the middle part will be found 17 inches.

DIRECTIONS to the FOUNDER.

The corf wheel, plate 1, fig. 6, ſhews both a ſide and end view; the outſide diameter of the model is $13\frac{1}{4}$ inches, and the weight when caſt is 14lb. 3oz.; the center hole $1\frac{1}{16}$ diameter, and the end of the nave meaſures 2 inches over, and $2\frac{1}{2}$ inches in length through the axis. The ſpokes, which are eight in number, are 9-16 of an inch thick in the middle, and $\frac{1}{4}$ at the edge, and meaſure 2 inches broad at the nave, and $1\frac{1}{4}$ at the rim; the rim is $1\frac{1}{2}$ inches broad on the trod or face, and 9-16 thick in the middle, and $\frac{1}{4}$ at the edge.

The buſhes, plate 1ſt, fig. 4, gives a front and ſide view of them; the oblong part meaſures 5 inches by $2\frac{3}{4}$, and is $\frac{3}{8}$ thick, with 4 counter ſunk rivet holes through, and the circular part which is 2 inches outſide diameter, with a hole through, bare $1\frac{1}{4}$ inch diameter, meaſures $1\frac{1}{4}$ inch through the hole, and muſt be *rounged* or *bored* to full $1\frac{1}{4}$ inch diameter, and the wheels muſt be rounged to $1\frac{1}{8}$ inch full.

ESTIMATE of the EXPENCE of the CORF.

6ſt. 7lb. of iron work, at 5s. per ſtone,..........................	1	12	6
Turning the axletrees, and rounging the wheels and buſhes,..	0	2	4
Ends and ſides, and ſawing them,..................................	0	4	0
3 bars in the bottom, and boxes for covering the wheels,......	0	1	1
Carpenter's work, and laths or flags for the bottom, and nails,	0	5	11
Wheels and buſhes 4ſt. 2lb. at 2d. per lb.........................	0	10	2
Nogs and boxes for mottys, or ſticks, to diſtinguiſh the Corf,..	0	0	6
TOTAL. £.	2	16	6

The corf, (plate 1, fig. 15, ſhews the ſide view,) is well adapted to ſhort work, or what is generally called *benks* or *boards*, and where the coals are in a great meaſure large, and the bed of coal not very thick.

DIMENSIONS.

Length of the outſide 42½ inches, breadth 31½ inches; inſide length 40½ inches, breadth 29, and height 19 inches, and ſtands 26 inches high upon the wheels, which are only 10 inches diameter. Contains neat meaſure, 21100 ſolid inches.

DIRECTIONS to the SAWERS and CARPENTER.

Every article that varies from the other corf, is in the ſides, which muſt be cut 44½ inches long, and 23 inches broad, the ends which are 33 inches long and 19 inches broad; the *Oak ſpendings* which are 32 inches long, and the flags for the bottom 40½ inches long.

The buſhes are 13¼ inches aſunder from center to center, and the mortiſe of the ſpendings do not come nearer the corf end than 3½ inches.

DIRECTIONS to the FOUNDER.

The Wheel of this corf, which is 10 inches high, has only 6 ſpokes in it, and has in every other part the ſtrength of the other wheel; its weight is 9¾lbs.

DIRECTIONS to the BLACKSMITH.

The axletree (plate 1, fig. 16, muſt be 2 feet 9 inches long, and muſt meaſure 4½ inches to the fixed boſs, which muſt be ½ an inch, thence to the edge of the *cotter hole*, which is ⅝ long, is 1 foot 10½ inches, and from the cotter hole to the end 4⅞ inches. The plate that lays along the bottom of the corf meaſures from the end to the center of the firſt hole 1¼ inch, thence to the ſecond 4¼ inches, and thence to the middle 15⅝ inches; which makes the plate 3 feet 6½ inches long; and ſhould be 2¾ inches broad, and full ⅛ of an inch thick, if the corves are intended to be hurried by horſes. The angular bottom plate is 5 feet 11¾ inches long, 2 inches broad, and bare ⅛ thick; the firſt hole meaſures 1¼ inch, thence to the ſecond hole is 2 feet 4⅝ inches. The plate that croſſes the end of the corf at the top, meaſures 4 feet 5 inches long, 2 inches broad, and bare ⅛ thick, which requires nothing but nail holes in it. The ſuſpending lug of the corf ſhould be 2 feet 3½ inches long, 2 inches broad, and ¼ thick; the holes below the ſuſpending hole may be about 3 inches, and the other hole within 1 inch of the end; the mortiſe plate is the ſame as that of the other corf, and the low corner plate may be 18 inches long, and a ſquare hole in it 4½ inches from the end. The expence of this corf is 2l. 13s. 6d. and its weight 2¾ cwt.

CAST

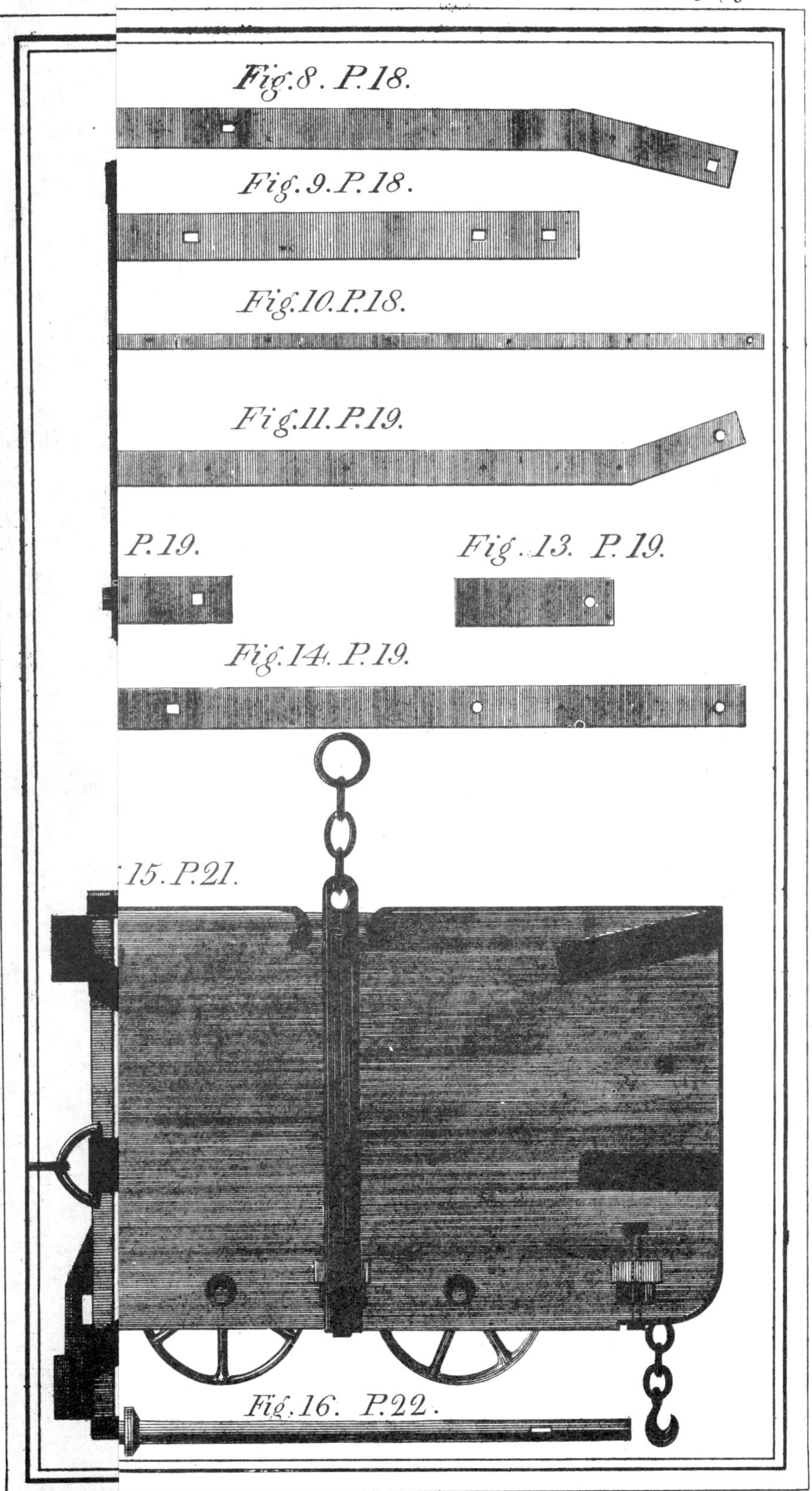

Engraved by Thos. Harris, Sheffield.

CAST IRON RAIL ROADS.

THE COMMON PLATE.

THE Plate of general uſe ſhewn plate 2, fig. 8, (which ſuits both ſides of the road) is 6 feet long, 3 inches broad on the trod, and $\frac{1}{2}$ an inch thick. The margin ſtands 2 inches high above the plate, and is $\frac{1}{2}$ an inch thick where it joins upon it, but is tapered to the top (which is rounded) to $\frac{3}{8}$ of an inch thick, for the convenience of moulding. There muſt be counter ſunk nail holes within 1 inch of each end, and the lugs for fixing the plate in the ſleeper may be $1\frac{3}{4}$ of an inch long, and meaſure when put on $4\frac{5}{8}$ inches broad over the bottom. One end of the common plate is ſhewn on an enlarged ſcale, fig. 9. The joiner muſt be particularly careful to make all his models $4\frac{5}{8}$ inches broad at the ends, as a want of attention to this, occaſions a great deal of trouble when the plates are laid down. The weight of this plate is from 47 to 50lb.

This rail plate is well adapted to the corves heretofore deſcribed, and *hurrying* or *putting* by horſes; and when greater burdens are neceſſary to be taken in each corf, the plates may be ſtrengthened by caſting them 4 or $4\frac{1}{2}$ feet long, and the margins may be raiſed $\frac{1}{2}$ an inch higher

in

in the middle, and tapered down to 2 inches at the ends; and if very great burdens are required, the metal may be made in general $\frac{1}{8}$ of an inch thicker; and on the contrary, if the corves are lighter than thoſe herein deſcribed, the plates may be made $\frac{1}{8}$ of an inch thiner.

THE SLEEPER.

Shewn fig. 10, on an enlarged ſcale, is 3 feet, 4 inches long for the wide corf heretofore deſcribed, and 3 feet $2\frac{1}{2}$ inches for the ſtraiter one; ſhould be ſawn out of Oak, $4\frac{1}{2}$ or 5 inches broad, by $2\frac{1}{2}$ inches thick, and the plate muſt be ſunk down 1 inch deep into the ſleeper, and the road muſt be laid down $22\frac{1}{2}$ inches wide, to ſuit the narrow corf, and 24 inches wide to ſuit the wide one, affording about $\frac{5}{8}$ or $\frac{3}{4}$ of an inch play in the corf wheels, which will be found quite enough in the ſtraight or eaſy bending roads, but for very quick turns the play requires to be $1\frac{1}{2}$ inches.

BROAD ENDED PLAIN PLATES.

Shewn fig. 1. The narrow end joins the common roads, and the broad end (1 inch broader than the other) joins the turn plates next explained.

PLAIN

PLAIN TURN PLATES.

Used for going round a turn, shewn fig. 2. The *trod* or *tread* of these plates are 4 inches broad, and forms a quadrant; and on account of the turn must be laid 1½ inches straiter than the straight road. The margin of the inside plate is drawn with a radius of 3 feet 2 inches, and that of the outside plate is drawn with a radius of 4 feet 11 inches, which will be found convenient for the narrow corf, and for the wider corf the outside margin must be drawn 1½ inch wider.

PLATES for turning into BENKS or BOARDS.

Shewn fig. 3. The *trod* in the straight part of these are 3½ inches broad, and the circular part 4 inches broad, and forms also a quadrant: a view of the figure will sufficiently explain it; and for turning into *benks*, where more room can be given, the breadth of the plates may be the same as the other, but the margin of the inside plate may be drawn with a 4 feet radius, and must form a quadrant also.

POINTER PLATES. Fig. 7.

Useful for taking the corves upon the roads; the length of these pointers may be about 4 feet 6 inches or 4 feet 9 inches; the pointer end of these are 5½ inches broad, which admits of them lying 2 inches clofer together at that end of them, and makes the corf pafs on with lefs friction.

PASS BYE PLATES. *Fig.* 12.

Ufeful for 2 horfes going contrary ways and paffing each other with a draught of corves. The models in general of thefe plates fhould be 4 inches broad on the *trod*, as they are obliged to be laid a little ftraiter than the common roads. A fight of the figure clearly reprefents the mode of turning out to the right hand, and paffing, which is done very fafely, and *without a fwitch rail*, as is required in common waggon ways. The firft pair of plates *(a)* which are 6 feet long continue ftraight for 4 feet, and are $7\frac{1}{2}$ inches broad at the broad end. The fecond length *(b)* is a double fet, the fhorteft of them muft be 7 feet 3 inches long. The third length *(c)* requires to be only 6 feet long. A view of the figure will convey a more clear idea than any explanation that can be given, obferving only, that the narrow part in the length *(c)* where the margins require to be double, on account of a part of the middle plate being without a margin, muft be only $2\frac{1}{8}$ inches wide; and the common length of plates *(d)* muft be as often repeated as to allow fufficient room for the draught of corves to ftand in the paffing.

ſing. To prevent the corf wheels running againſt the point, *(e)* it is neceſſary to raiſe that ſide of the road 2 or 3 inches higher, for the length of 3 or 4 pairs of plates; and ſuppoſing a branch of road is required to be made to a new pit, or any particular place, one end of the above deſcribed *paſs bye* (viz.) *(a) (b)* and *(c)* will accommodate ſuch purpoſe, taking out the plate, *(f)* and ſhortening the two next adjoining plates, and introducing the long plate fig. 13, with the *ſwitch rail (g)* upon it, which works upon a pin, to turn the corves occaſionally out of the direct line of the road; the part *(h)* being a ſtop to prevent the *ſwitch (g)* from flying out too far.

PLATES for the MOVING DOUBLE ROAD, to convey large coals upon the GROUND, for STACKING them.

Fig. 4. ſhews the platform of the firſt pair of plates; as the planking lies nearly level, and the ground deſcending, the pointer part (*a*) which lies upon the planking, muſt bend a little downwards, to make it lie ſolid; a part of theſe plates are alſo ſhewn, plate 3, fig. 1. For 5 inches long, it is let into the planking, and the ſmall holes take ¾ inch diameter pins of iron, to hold the plates in their places. The ſecond length of plates for the *Jinney*, which are thoſe chiefly uſed, fig. 5 ſhews the platform of one pair, and fig. 11 ſhews a ſide view of the margins.

Fig.

Fig. 14 ſhews the oak *ſleeper* for the double road with 4 pins, each $\frac{1}{2}$ an inch diameter ſtanding up, which goes into the holes in the margin, and holds the plates at proper diſtances. The end of theſe plates muſt be taken a little under the ſquare, that they may move about without injuring the joints.

PLATES for EASY TURNS in the ROADS.

Fig. 6 ſhews a pair of plates which bend .18 parts of an inch in the middle, are drawn with a radius of 100 yards, and accommodate a turn of 9 inches in every 5 yards, the road being 2 feet wide. The long ſide of the long plate meaſures 6 feet, and the ſhort ſide of it 5 feet 11.94 inches. The long ſide of the ſhorter plate is 5 feet 11.46 inches, and the ſhort ſide 5 feet 11.37 inches.

If a few of theſe turning plates are wanted, this great accuracy is not ſo important, but if a great length of them is required in a road, if this accuracy is not attended to, the joints will be open, and a great deal of friction will be unneceſſarily added. It muſt be further obſerved, that the long ſide of the roads in all turns, muſt be raiſed 2 or $2\frac{1}{2}$ inches higher than the inſide, by which means the gravity of the loaded corf takes the friction of the wheels from the inſide margin of the road.

If

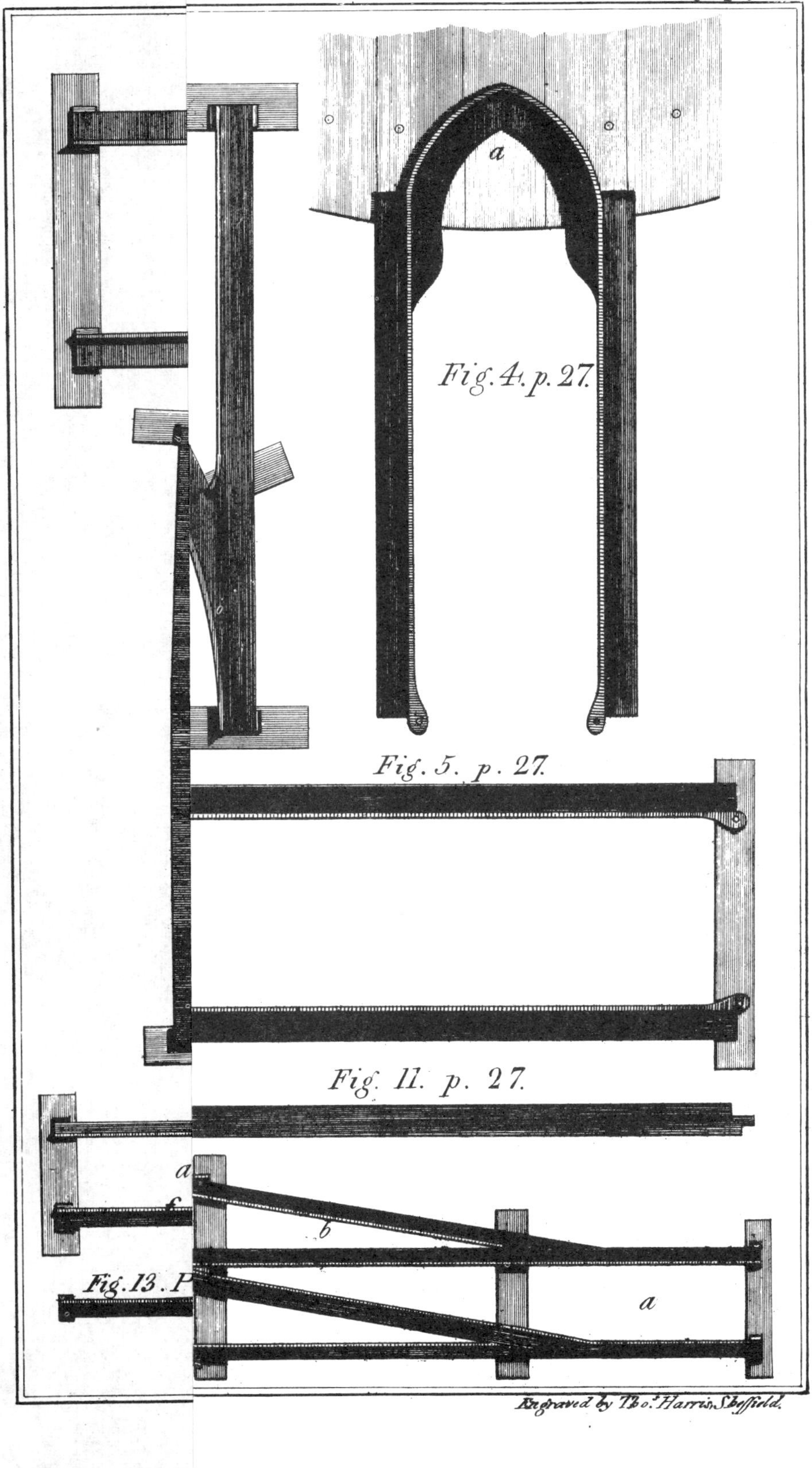

Engraved by Thos. Harris, Sheffield.

If a road is wanted to accommodate a turn of 18 inches in 5 yards (equal to the ſweep of a radius of 50 yards,) theſe plates muſt bend *.37 of an inch in the middle; the long ſide of the outer plate muſt be 6 feet, and the ſhort ſide of it 5 feet 11.88 inches. The long ſide of the inner plate will be 5 feet 10.93 inches, and the ſhort ſide of it 5 feet 10.81 inches. For a bend of 12 inches in every 5 yards, (equal to the ſweep of a radius of 75 yards,) the plates muſt bend .24 of an inch in the middle. The long ſide of the outer plate will be 6 feet, and the ſhort ſide of it 5 feet 11.92 inches. The long ſide of the inner plate will be 5 feet 11.29 inches, and the ſhort ſide of it 5 feet 11.21 inches. For a bend of 6 inches in every 5 yards, (equal to the ſweep of a radius of 150 yards,) the plates muſt bend .12 of an inch in the middle; the long ſide of the outer plate will be 6 feet, and the ſhort ſide of it 5 feet 11.96 inches; the long ſide of the inner plate will be 5 feet 11.64 inches, and the ſhort ſide of it 5 feet 11.6 inches.

To explain more clearly what is here meant by a certain deſcription of *turn* (viz.) 6, 9, 12 or 18 inches in every 5 yards, ſee plate 2d, fig. 15, where (*a*) to (*b*,) (*b*) to (*c*,) and (*c*) to (*d*) meaſure 5 yards, and (*e*) to (*b*,) (*e*) to (*c*,) and (*e*) to (*d*) meaſure 6, 9, 12, or 18 inches as neceſſity requires; by which mode of ſetting out you attain a regular *turn*.

G *JINNEY*

* 37 Hundred parts of an inch.

JINNEY

FOR CONVEYING THE CORVES ABOVEGROUND.

PLATE 3, fig. 5, gives a front view of the *jinney*, the barrel (*a*) of which is 4 feet 6 inches diameter. Fig. 10 ſhews the ſide view of it, and fig. 1, ſhews the platform of it, which turns upon a pin in the center, and points to any required direction. A part of the planking for the corves to turn upon, and the points of the plates (*b b*) is alſo ſhewn. The ground on which the coals are ſtacked has a deſcent of about 3 inches in the yard from the *jinney*, and the momentum of the full corves going down the inclined plane, with the aſſiſtance of the communicating ropes, takes the empty corves back to the *jinney*. If power is wanted, there is a handle (*c*) to aſſiſt the *jinney*, and if it has too much velocity, there is a *brake* (*d d*) to retard its progreſs.

JINNEY for CONVEYING the CORVES UNDERGROUND.

Plate 3d, fig. 7, gives a front view of the *jinney*; the two rope barrels (*e e*) are fixed in two inclining *board gates*, on which the corves paſs, which are divided by a pillar of ſolid coal 4 yards thick. The ropes communicate round the barrel, and work upon the ſame principle as the *jinney* above ground, before deſcribed, and the narrow wheel (*f*) at one ſide of the rope wheel, is to retard its motion by the application of a *brake*.

JINNEYS *for* CONVEYING *the* CORVES.

ESTIMATE of the EXPENCE of the JINNEY ABOVEGROUND, and SCANTLING of the WOOD.

	£.	S.	D.
40 fuperficial yards of planking, 1½ inches thick, and fleepers for the foundation,	4	6	8
1 Sole tree 7 feet long, and 6½ inches by 5½,	0	3	6
2 *Uprights*, 6 feet long and 2½ inches fquare,	0	3	0
2 Side braces, 3 feet 2 inches long, and 3 inches by 2½,	0	1	0
2 Sole trees, 6 feet long, and 4½ inches fquare,	0	3	6
4 Braces, 4 feet 8 inches long, and 2¾ by 2½,	0	2	0
1 Crown tree, 4 feet 10 inches long, and 3½ inches fquare,	0	1	0
1 Axletree, 3 feet 2 inches long, and 5½ inches fquare,	0	1	6
2 Crofs *fheths*, 4 feet long, and 3½ inches fquare,	0	1	6
8 Arms, 4 feet 3 inches long, and 3¼ by 2½,	0	4	0
2 Cribs, 3 inches by 2½, and 1 fhroud 7 inches by 1,	0	5	8
1 Shroud for the middle, 2¾ by 1, and 1 ditto for the *Brake*, 2¾ inches fquare,	0	4	0
Boarding the face of the wheel,	0	4	6
The brake with 2 pieces, each 4 feet long, and 4 inches by 2½,	0	1	0
1 Piece 2 feet long, and 4 inches by 3,	0	0	6
Iron work,	0	15	0
Carpenter's work and fawing,	2	2	0
TOTAL AMOUNT.	£. 9	0	4

ESTIMATE of the EXPENCE of the JINNEY UNDERGROUND.

	£.	S.	D.
1 Axletree 20½ feet long and 3 inches fquare,	0	10	0
16 Arms 4 feet long each, and 3 inches fquare,	0	6	0
4 Cribs, 3 inches fquare each, and 1 ditto 3½ by 3, and brake,	0	13	0
Boarding on the face 3 inches broad, and 2½ inches afunder,	0	6	0
2 Punches or props for the jinney to work in,	0	3	0
Iron work,	0	9	0
Carpenter's work and fixing,	1	10	6
TOTAL AMOUNT.	£. 3	17	6

MACHINE

FOR OPENING DOORS UNDERGROUND.

PLATE 3, and fig. 2, ſhews the ſide view of the machine, one of which is required on each ſide of the door, at about $4\frac{1}{2}$ yards diſtance from it, to allow ſpace for a horſe, and room for the door to open. *(g)* is a moving *lever*, and *(h)* is a ſmall weight which holds the *lever* in a proper ſtate for the corves to catch it; and the door is hung with a clap ſufficient to make it ſhut of itſelf. Fig. 9, gives the platform of the machine, and *(i)* ſhews the platform of the door. When the corves advance forward, the corner of the firſt of them runs againſt the *lever (g,)* and puſhes it forward, until it has performed a ſufficient ſtroke at the other end of the *lever* to open the door, by means of the communicating rope *(k)*. The Corves as they paſs forward, hold the *lever* in its poſition, and when they have paſſed the *lever*, the *board* (*l*) which forms a ſegment of a circle, and is nailed upon the door) prevents the corves from catching it.

ESTIMATE of the EXPENCE of the MACHINE & Scantling of the WOOD.

1 Upright 5 feet long and 7 inches ſquare,..................	0	3	0
2 Arms, 3 feet 3 inches long, and 3 inches by $2\frac{3}{4}$,............	0	1	0
Lever 3 feet 1 inch long, and 2 inches thick; 9 inches broad at one end and 4 at the other,..........................	0	0	6
Wheel, 9 inches diameter and 2 inches thick,................	0	0	6
Sheaves and fixing parts 3s. 8d.....iron work, 5s................	0	8	8
Carpenter's work and fixing 6s.....Caſt iron weight and buſhes, 2s 3d. and Rope 2s. 6d..........................	0	10	9
EXPENCE of the MACHINE on one ſide of the DOOR. £	1	4	5

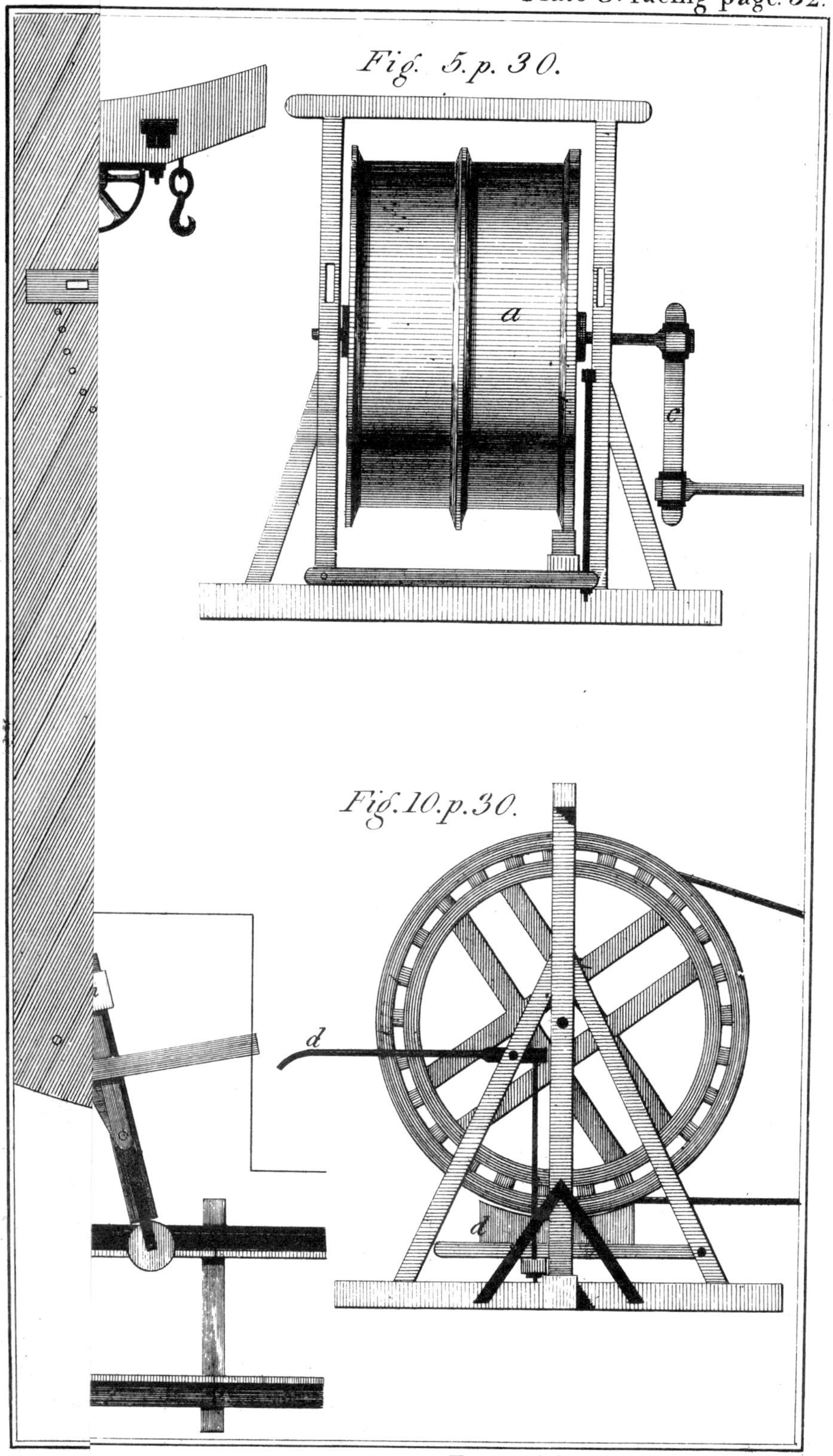

Engraved, by Thos. Harris, Sheffield.

MACHINES
FOR DRAWING COALS.

THE machines for drawing coals, alluded to in the introduction for the conveying of coals underground, being ſo great an acquiſition to collieries in general, eſpecially where the depths are great, the ſubject ſhould not be paſſed over without a further explanation. What is here meant by machines, are the different modes invented for the purpoſe of drawing coals or other minerals out of pits, without the uſe of *gins*, or *jack rolls* wrought by hand.

Were I to enumerate or explain the many fruitleſs attempts, together with the few ſucceſsful ones of this nature, which I have ſeen or heard of, the taſk would be arduous indeed; but a ſhort explanation of the principles of theſe in practice, which have fallen within the compaſs of my obſervation, may be uſeful to ſome of my readers.

The moſt ancient machine in my knowledge, now in uſe, is that invented by MENZEY, but there are few ſituations that afford the requiſites neceſſary to that invention. A ſtream of water with a waterfall of about half the depth of the pit is neceſſary, if any buſineſs of *conſequence* muſt be done. Its conſtruction conſiſts of two rope wheels fixed upon one horizontal axis, which are ſo proportioned to the depths of the water pit and coal pit, as to reach the ſeparate depths of the pits, by the ſame revolutions; and the power applied is a tub of water large enough to overbalance the weight to be drawn.

The ſecond is the common machine, greatly in uſe in the neighbourhood of Newcaſtle-upon-Tyne, the conſtruction of which is, a water wheel and a rope wheel upon one horizontal axis; and the power is a ſtream of water ſufficient to overbalance the weight to be drawn. The method of obtaining this ſtream of water in all the collieries in the neighbourhood of Newcaſtle-upon-Tyne and Sunderland, where there are, I preſume, no leſs than 30 or 40 in number, is a Fire Engine placed by the ſide of the machine, which raiſes the water alternately to the top of the wheel; but in two collieries where I have adopted them, the ſcheme is more advantageous than thoſe at Newcaſtle, being able to do without a Fire Engine erected ſolely for that purpoſe; in the winter ſeaſon when water is plentiful, and the engines are generally ſufficiently employed with draining the collieries, we have the aid of adjoining brooks which do our buſineſs; and in the ſummer ſeaſon, our engines are ſo conſtructed, as to apply a part of their power to raiſe the water to the top of the wheel.

The

The third and laſt, of any importance in my knowledge, is the Fire Engine immediately applied to the act of drawing without the aid of a water wheel, of which there are ſundry kinds. One is the invention of Meſſrs. BOLTON and WATTS, another of Mr. CAMERON, which does not differ much in principle, and a third is the common fire engine, which was firſt reduced to practice by the Engineers of Colebrook Dale collectively.

Where the ſituation is ſuitable, the machine invented by MENZEY has ſimplicity to recommend it, and by drawing two corves at a pull, is capable of doing a great deal of buſineſs. The common machines with a water wheel and engine, have been chiefly built before the third plan of a fire engine applying its power immediately, was made manageable and uſeful; and as it does not require more than half the power when the engine alone is applied, and the original expence being little more than one third, we can have no difficulty in condemning the further introduction of the water wheel, excepting where a brook can be conveniently had to do the buſineſs.

It ſeems ſomewhat extraordinary, that the drawing by machines has not yet made greater progreſs in the ſouthern parts of this kingdom, which I can ſee no objection to, but in two points, and thoſe in few caſes only amount

to

a real difficulty; the firſt is the extra hurrying, putting or conveying underground, and the ſecond is the difficulty of drawing coals up the ſhaft in ſuch corves as are capable of being introduced in theſe parts, where the coals are large, with the quick pace of a machine. In anſwer to the firſt objection, I can only obſerve, that the inſtructions, here laid down for making corves and rail roads, in ſituations where the bed of coal affords height for a horſe, or poney 12 hands high, will enable any proprietor or manager of works, to take the coals three or four times the diſtance, without incurring any objectionable proportion of comparative expence; and even where men or boys only can be introduced, the quantity conveyed, and the diſtance, may be greatly increaſed; and with reſpect to drawing up the ſhafts at a quick pace; being myſelf the patentee for the invention of conductors to prevent damage to the corves and ſhafts, I will only recommend it to the intereſted public, to take a view of the methods now eſtabliſhed at ſundry collieries near Sheffield, Barnſley and Leeds, and let them judge for themſelves. Theſe conductors are nothing more than two or three *upright rods* of deal 4 inches by 3, braged upon oppoſite ſides of the pit, forming mortiſes or channels, by which the corves are conducted, being ſuſpended upon croſs-bars with rollers at their ends, which run within the mortiſes.

THE

THE FIRE ENGINE.

THE nature and principles of the common fire or ſteam engine having been heretofore ſet forth and demonſtrated by much abler pens, and being now, it is preſumed, very well underſtood, I ſhall be ſilent upon that head; but when the principles are perfectly known, there remains a great deal to be done to proportion the various ſizes and ſtrengths of the materials, and arrange them in a proper manner.

If we give a glance through this kingdom we behold with aſtoniſhment, fire engines, even at the preſent day, that do not more than half the buſineſs they would be capable of, if properly conſtructed. The various applications to which the fire engines under my care are adapted, have afforded me the opportunity of making ſeveral obſervations in this moſt uſeful of all machines, which I conceive to be of too great importance, to paſs unnoticed, and have been ſo fortunate as to hit upon ſome deviations from the general rule in certain engines I have erected, which have produced an effect far exceeding my expectations, and which I flatter myſelf will be deemed worthy the attention of engineers, as I can inform them, I have obtained a conſiderable addition of power, without any increaſe of fuel.

It is a well known principle in ſteam engines, that the more perfect the vacuum can be made in the cylinder, the greater power you will obtain; and as the vacuum is obtained by *ſteam*, and a *jet* of cold water, it is obvious that the higher the jack head ciſtern is placed that commands the injection, the more minute will be the diviſion, and the more rapid the diſperſion of the particles of water which condenſe, and in conſequence, the more perfect the condenſation. What then can we expect of thoſe engines whoſe ciſterns are placed at the height only of 12 or 14 feet above the top of the cylinder, compared with the ciſterns of the engines above alluded to, which are placed about 36 feet above the top of the cylinder, the conſequence of which is, that we obtain a more perfect vacuum than thoſe cylinders can obtain with a low fixed ciſtern where the injection riſes faintly.

In order that my readers may compare the operation of a ſteam engine above alluded to with others now in uſe, and judge for themſelves of their reſpective merits, I muſt inform them, that the diameter of the cylinder is 61 inches, with two boilers each 14½ feet diameter, which conſume 10 cwt. of ſmall coal or ſleck in an hour; and works a ſtroke 8 feet 6 inches long. For a particular plan of it, ſee plate 4, in which fig. 1 gives a ſide view, fig. 3 a front view, and fig. 8, the platform. The front and

and ſide view of its working *geer* is alſo ſhewn fig. 2, on an enlarged ſcale. This engine lifts a ſet of pumps of a 13 inches working barrel 24½ fathoms, a 13⅛ inches ſet 23½ fathoms, a 15 inches ſet 7 feet 4 inches from the center of the beam, (which is 25 feet long) 5 fathoms, a 15¼ inches ſet 6 feet 2½ inches from the center of the beam 5¼ fathoms, and the *jack head ſet*, which is 9 inches, at 8 feet from the center of the beam, 10 fathoms; all of which when proportioned to the end of the beam, makes it appear that the engine works to 7lb. per ſquare inch upon the piſton.

When it better ſuits the convenience of the work, we draw the 13 inches ſet of pumps only 45¼ fathoms high, and work a 7 inches ſet alſo the ſame height, and take off the ſhort *lifts*, which brings the engine to 7¼lb. preſſure upon the ſquare inch; but the former ſtatement is the general ſituation of the engine, in which it performs 12 ſtrokes each, 8½ feet neat, per minute, without laying any unreaſonable burden upon the boilers.

Perhaps there is not a cylinder of 61 inches diameter upon the common conſtruction that works to 7lb. to the inch, and performs above 10 ſtrokes per minute 7 feet long each, with a moderate quantity of ſteam, except thoſe built upon the ſame conſtruction, ſince the engine alluded to was erected, and this it appears is doing near half as much more buſineſs.

For

For the information of ſuch mechanicks, as have not had an opportunity of making experiments on the *power*, or *burden* upon the piſton, by which the engine will do the moſt execution, I muſt inform them; that when I annexed the ſmall *lift* of pumps 7 inches diameter above mentioned, which raiſed the burden of the cylinder to 8½lbs. per ſquare inch, upon the piſton, it would not, notwithſtanding the utmoſt efforts were exerted with the boilers, perform above 9 ſtrokes per minute, each ſtroke only 8 feet long, which is far inferior to the execution, when burthened to 7lbs. per ſquare inch only.—I alſo gave this engine a trial with a burthen of 6.1lbs. per ſquare inch, in which ſtate it performed more real execution than when burthened to 8½lbs. per inch, and ſomewhat leſs than when burthened to 7lbs. per ſquare inch; but as the two material parts of the engine (viz.) the regulator and injection by frequent working can ſeldom be kept perfect, I would recommend that no engine ſhould be laid to a higher burthen than 6¼ or 6½lbs. per ſquare inch.

As every article of the engine here alluded to is upon the common conſtruction, excepting the raiſing of the *jack head* ciſtern to a good height, we muſt inevitably conclude that the extra merits it is poſſeſſed of, muſt ariſe from that cauſe, together with a judicious arrangement and proportioning of its conſtituent parts. In

In juftice to the public, I cannot conclude this fubject without obferving, that the method of fitting up Engines with valves, is in my opinion preferable to the common regulator and injection cock, as heretofore defcribed, being much lefs liable to be out of order, and more eafy to repair; and the annexed plan, plate 4, fig. 7, fhews a fide view of the low part of the cylinder 48 inches diameter, and working *geer*, with a fteam cheft upon a good conftruction, *(a)* being the fteam valve, *(b)* the injection valve, *(c)* the hotwell, and fink pipe, and *(d)* the *plug* to work the irons; and fig. 6 fhews a front view of the fame.

What I have now chiefly to offer in regard to the Fire Engine, is to give the ftrength and proportion of all the parts, and of all dimenfions of engines, in the beft manner I am able, from the refult of extenfive practice, and accurate obfervation. Thefe dimenfions apply generally to the common engine fitted up with a regulator and injection cock, and the alterations that take place in fitting up the engine with a fteam cheft and valves, are explained in the following inftructions, given under the heads of fire engine materials proportioned, and directions for building the fire engine.

It needs no elucidation to enable us to conclude, that

the weight or preſſure at the bottom of the pumps is much greater than at the top, for which reaſon I have endeavoured to proportion the thickneſs of the metal of the pumps to the burthen they have to ſuſtain. It will alſo be obſerved, that all the directions hereafter given relating to the fire engine, ſuppoſe the ſtroke to be 9 feet long, and to work 8½ feet neat in common; and I have in my directions for engine houſes, endeavoured to keep them as ſmall as can conveniently be diſpenſed with, to avoid ſuperfluity in expence, and ſhorten the timber; and the mode of fixing cylinders upon pillars, and ſcrewing them down by *under croſs beams*, I find by experience to have a better effect, than any ſtrength or quantity of *croſs beams* in the way of hanging the cylinder, that can poſſibly be introduced.

The annexed table given in the pages 76 to 81 incluſive, ſhews the diameter of the cylinder ſuitable to ſundry depths of pits and diameters of pumps, and the water ſuch engine will draw in a minute and in an hour, performing any given number of 6 feet ſtrokes. For example a 10 inch *bore* for a pit 50 fathoms deep requires a cylinder 43.1 inches diameter; draws 20 gallons of water at a 6 foot ſtroke, and by working 12 ſtrokes each 6 feet long per minute, draws 240 gallons of water per minute, or 228 hogſheads and 36 gallons in an hour, and of

courſe

courſe if the engine works a 9 feet ſtroke, it will draw juſt half as much more water as the table ſets forth. But the cylinder muſt always be made larger than the direction given in the tables, as 6½lbs. per inch is a ſufficient burthen for an engine to work to, and the table is calculated at 7lbs. per inch, and a further allowance muſt alſo be made for the *jack head*, and in calculating thoſe allowances, the tables given in pages 82 and 83, will be uſeful.

FIRE ENGINE MATERIALS
PROPORTIONED.

Boilers proportioned to Cylinders, ſee page 70.
Conſtruction of Boilers, ſee plate 4, fig. 3.

I HAVE in the courſe of my practice, tried both *flange boilers*, and plain ſided ones, with *concave* and *convex* bottoms, but muſt give greatly the preference to the plain ſided ones with concave bottoms; ſee the table of directions for making or planning the boilers, page 89, where the *thickneſs* of the *plates* and weight thereof are alſo given; and I would adviſe, that no boiler ſhould be conſtructed of a larger diameter than 17 feet. The annexed table page 88, gives the exact length, and breadth of the plates

plates at each end, for boilers of all dimenſions, and as the top plates of a boiler require to ſwell a little in the middle, ſee the table page 91, which gives the intermediate breadths of them in ſundry parts.

DIRECTIONS to the BLACKSMITH for making BOILERS.

The rivet holes of the boiler bottoms ſhould be full ½ an inch diameter, as far as the top of the *flue plates*, and may be 2 inches aſunder from center to center of the holes, and for the top of the boiler, they may be ½ an inch diameter, and 1¾ inch aſunder from center to center. The *over lap* of the plates ſhould be ¾ of an inch on each ſide of the center of the *rabbit holes*, which makes the joints 1½ inch broad in the double plate.

The proper HEIGHT of WATER in BOILERS.

The Water ſhould always ſtand not leſs than 2 or 3 inches below the *joint* at the top of the *flue plates*, and the *gauge cock* ſhould go down to the top of the water.

DIRECTIONS for FIXING the BOILERS.

The *plug floor* in all the common engines falls 17½ inches below the top of the boiler, and in the valve engine it falls 2 feet 1 inch below, and for the height of the boilers

Fig. 7. p. 41.

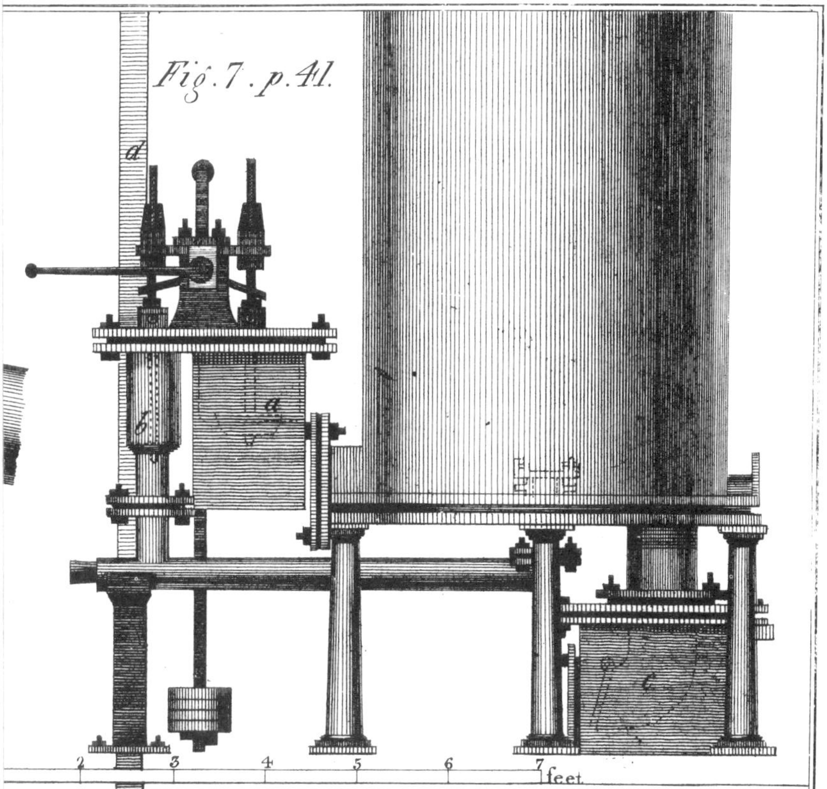

Fig. 8. p. 38.

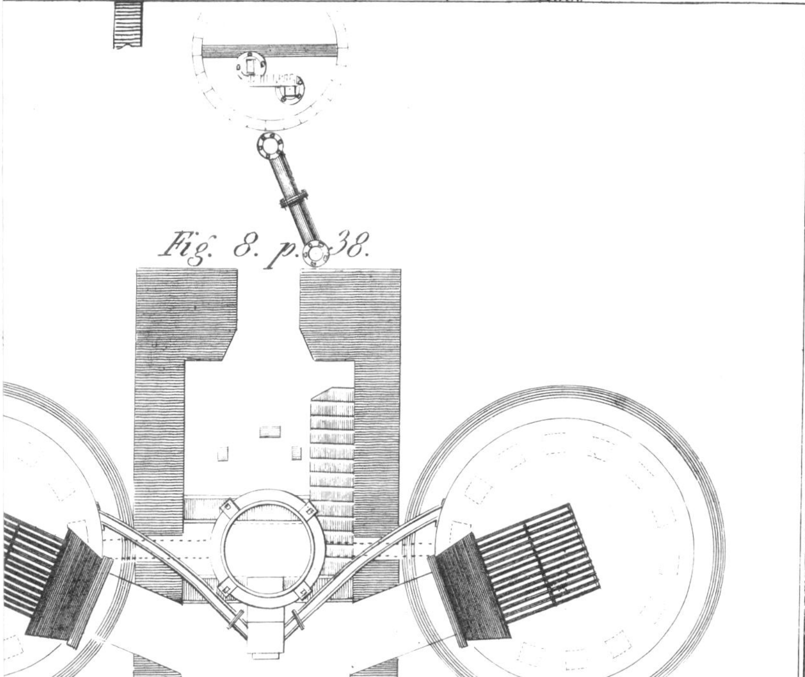

Engraved by Thos. Harris, Sheffield.

boilers ſee page 89; and the upper ſide of the *grate bars* in all engines muſt be 13½ inches below the *laggon* or loweſt part of the boiler, which *laggon* ſtands upon pillars about 15 inches ſquare and 15 inches aſunder.

The *center* of the boiler of the common engine muſt be placed at right angles from the center of the cylinder, and of the valve engine that ſtands at right angles from the *ſteam cheſt*, and the boiler muſt be placed at ſuch a diſtance from the outſide of the engine-houſe as to leave a ſpace of 19 inches, which allows 9 inches for *flue*, and 10 inches for *brick work*.

The height from the *aſh hole* to the top of the *grate bars* ſhould be 2 feet 9 inches, and the foundation ſhould be ſunk 5 or 6 inches lower, for pitching.

The *aſh hole* of the boiler ſhould extend 1 foot beyond the center of the boiler, and the length and width of the *aſh holes* before they are contracted for the fixing of the door, are given for boilers of every ſize, page 75.

The mode of fixing the boiler upon brick pillars, and incloſing it with a circular wall 10 inches thick, as high as the *womb* of the boiler, and 5 inches thick above, in the ſhape of a bottle, I do conceive from 7 years practice, to be a very good method, the heat being admitted to the boiler in a very impartial manner, which not only preſerves the boiler, but the grate bars alſo.

CAST IRON PLATES laid over ASH HOLE, plate 5, *fig.* 22.

The top ſide of theſe plates lies level with the grate bars and door frame, and makes up the ſpace between them; and ſhould be caſt 2 inches thick in the middle, and $1\frac{1}{2}$ at the ends. For particular dimenſions of theſe plates, ſee table, page 75.

The plate that lies over the *door frame* againſt the *laggon* of the boiler, muſt be hollowed out as ſhewn plate 4, fig. 13, with the *radius* of the boiler at the bottom of the *flue plates*, given in the table, page 89; the breadth of the broadeſt part of theſe plates may be about 13 inches, and the length of them may be about the medium length of the long and ſhort ſide of the plates above deſcribed, and ſhould be caſt about 2 inches thick.

BUCKETS of CAST IRON, plate 5, *fig.* 36, 37, 38.

All *buckets* above 9 inches diameter, to be caſt $1\frac{1}{4}$ inch leſs than the *working barrel* when turned, and ſmaller ones about $\frac{3}{4}$ or 1 inch leſs, and thoſe that exceed 14 inches diameter ſhould be raiſed a little on the ſides, for the eaſe of the *lids* opening and ſhutting, ſee plate 5, fig. 39, 40, 41, and the ſides of the *bucket* ſhould taper about $\frac{1}{8}$ of an inch in every inch deep. For the depth of them and the ſtrength of the metal ſee page 74, and if caſt in braſs, may do a little thinner.

CLACK

CLACK, and BUCKET SHANKS.

For the ſtrength of the malleable iron, ſee table page 74.

CLACKS.

To be caſt by the directions given for the *buckets*, excepting only, that the diameter muſt be $\frac{1}{2}$ an inch leſs than the *working barrel.* The ſtrength of the hoop of malleable iron for holding on the leather upon the *bucket* and *clack* is ſhewn table page 74.

CATCH PINS.

To fix in the heads of the *regulator beam,* are made of malleable iron; ſee the length and ſtrength of them p. 71.

CYLINDERS. See plate 5, *fig.* 1.

For the common engine, require the length for a 9 feet ſtroke to be 10 feet, and muſt be *bell mouthed* $\frac{1}{4}$ of an inch on each ſide; the *cup ring* to be placed 3 inches from the top, ſhould be 2 inches broad and $\frac{3}{4}$ deep; to have 4 *lugs* for hanging it, fixed at 3 feet 4 inches from the top of the cylinder, 2 of which ſtand on each ſide of the *cylinder* oppoſite each other, as dotted upon the platform of the cylinder bottom, fig. 5, with a hole in *each lug* 1$\frac{1}{2}$ inch ſquare. A 70 inch cylinder requires *theſe lugs* to be 3 inches deep, a 60 ſhould be 2$\frac{1}{2}$ deep, a 50 ſhould be 2 deep, a 40 ſhould be 1 ¾ deep, and a 30 ſhould be 1½ inch deep. The metal of a 56 inches cylinder and upwards, to be left 1⅛ inch thick, when bored; of cylinders from 46 to 55 may be 1 inch thick, and of 45 inches and under to be left ⅞ thick. The *flanges* of all cylinders as high as 40 inches diameter, ſhould be 3½ inches broad, and all above that ſize ſhould be 4 inches broad. The thickneſs of the *flange* of a 56 inches cylinder ſhould be 1¾ inch, for a 50 ſhould be 1½ thick, for a 40 to be 1¼ thick, and for a 30 to be 1⅛ thick. The holes in the *cylinder flange* may run 9 or 10 inches aſunder from center to center, and ſhould be for all cylinders under 40 inches diameter 1¼ ſquare, and all above 1½ ſquare. The cylinder for the *valve engine* muſt be 10 feet 6 inches long, and admits the ſteam by an oblong hole, as cloſe as poſſible to the *flange*, and the oblong pipe projects out 5 inches, with a *flange* upon it for the *ſteam cheſt* to fix to, as appears plate 4, fig. 7, and muſt have a ſniſt pipe which falls fair between the *lugs*, and ſtands oppoſite to the *ſteam pipe*.

CYLINDER BOTTOMS.

For the common engine ſee the platform plate 5, fig. 5, and ſide view fig. 2 and 3, which require to be caſt 12 inches deep, including the *flange* and *bottom*. The pipe in the center to admit the ſteam in a 70 inch cylinder (which

(which ſtands in all caſes 3 inches above the bottom and 9 inches below it) ſhould be 13 inches diameter; for a 60 12 inches, for a fifty 10½, for a forty 9¼, and for a 30, ſhould be 8 inches inſide diameter. The metal of the ſides to be the ſame ſtrength as the cylinders, and that of the *bottom* and *flange* to be the ſame thickneſs as the flange of the cylinders above deſcribed. The ſink pipe for all cylinders above 40 inches diameter to be placed *fair* under the *ſnifting*, and nearly cloſe to the ſide of the cylinder, projecting 4½ inches below the bottom, including the *flange*, and to be the ſame diameter as the ſink pipe hereafter deſcribed, page 63; but for cylinders under 40 inches diameter it muſt ſlope away from the bottom, as deſcribed in plate 5, fig. 33, and requires a ſink pipe of a particular deſcription ſhewn fig. 34. The *ſnift pipe* ſtands nearly oppoſite the injection, and projects 5 or 6 inches, and ſhould be for all cylinders, 4½ inches inſide diameter. The injection pipe muſt not be fixed *fair* in the middle of the cylinder, (or half way between the *lugs*,) but muſt be fixed 5 inches to the left hand, (looking towards the cylinder,) and muſt alſo be laid down to the *flange* as cloſe as poſſible; the inſide diameters of theſe pipes are given page 72. There muſt be 4 *lugs* caſt upon the bottom with a hole in each lug 1¼ inch diameter, (to take the bolts incloſed in the pillars for the purpoſe of ſcrewing down,)

down,) for all cylinders 45 inches diameter and upwards, which *lugs* muſt not be caſt to ſtand under the *hanging lugs* of the cylinder, but to fall juſt half way between them, the thickneſs of them to be the ſame as the *hanging lugs*, and only 6 inches ſquare. The cylinders under 45 inches diameter require only 2 lugs and pillars, but cannot be fixed ſquare of the houſe, without interfering with the communicating pipes, and the ſtairs or ſteps leading down to the *aſh hole*; they muſt therefore be fixed in an *angular ſtate*, which angle muſt be juſt 26 degrees to the right hand of the center of the communicating pipe, and taken from the center of the cylinder; in which poſition the *under croſs beam* will alſo lie for ſcrewing down the cylinder, and leave a road down to the aſh place. The croſs beams ſhould be *oak*, and a 15 inches brick wall ſhould be taken up under them, to make them ſolid; but where the *angular beam* is put in, two brick pillars will be ſufficient to ſupport the *croſs beam* if fixed in their proper places.

The *cylinder bottom* for the *valve engine* of every ſize is caſt ſquare, and requires no ſides to it, has a *ſink pipe* and *injection hole*, but without a *ſteam pipe*.

CROSS BEAMS under the PILLARS.

To ſcrew down the cylinder above mentioned, muſt lie 1 foot 7 inches below the *plug floor* in the common engine, but in the valve engine theſe *beams* form a part of the *plug floor*. See their *ſcantlings*, &c. page 70. *CISTERNS.*

CISTERNS.

For the *jack head* that ſtands upon the top of the engine houſe, ſee their particular dimenſions page 70.

COMMUNICATING PIPES, plate 5, *fig.* 24.

For the common engine, the length to the extreme point of the *flange* is given page 71, and the metal of all ſhould be $\frac{7}{8}$ of an inch thick, and that of the *flanges* 1$\frac{1}{8}$ thick, and 4 inches broad. The angle of the *flanged end* of the communicating pipes of all, to be 35 degrees, and muſt be hollowed out by a circle of the radius of the boilers, and the valve for letting out the ſteam may be fixed near the *flange*. A *caſt iron ring* about 14 inches long, and metal 1 inch thick, for ſecuring the joint next the receiver, is much preferable to *a lap joint*, and ſhould be made large enough to allow $\frac{3}{4}$ of an inch for *wedging*.

For the *valve engine* they are required to be about 2 feet longer than thoſe for the common engine, with a *flange* at the end to join to the *ſteam cheſt*, and I would adviſe them to be caſt in 2 parts, with a *ring* to make the joint good, as deſcribed above for the common engine.

CYLINDER BEAMS.

To hang the cylinders upon, for particulars ſee page 70.

CHIMNEY PIPES.

For the *boilers*, ſee the diameters of them page 75.

DOOR FRAMES, plate 5, *fig.* 10.

For all the *boilers*, the inſide height ſhould be 13½ inches, and width 19 inches, the *uprights* (with *holes* in them for the *hooks* 1¼ ſquare) to be 4 inches by 5, and the length of the top and bottom part about 5 feet, and 4 inches ſquare.

ENGINE HOUSES.

The inſide dimenſions, and ſtrength of the walls is ſhewn page 70, and the *ground plan* of an engine houſe is ſhewn plate 4, fig. 8, and for the number of bricks, maſons bill, and quantity of lime and ſand uſed in all engine houſes, in medium ſituations, ſee page 70.

FRAMES in the ENGINE HOUSES.

Laid in the walls at the height of the *cylinder beams* and *regulator beam* to ſtrengthen them, ſee page 71, and plate 4, fig. 5.

GUDGEONS, plate 5, *fig.* 18, 19.

Proper for the *regulator beam* to work upon, with a little *curve*, and are ſuppoſed to be ſunk down 1 inch into the *beam*, which requires no other faſtening than 8 *iron ſcrews* 12 or 14 inches long, ſcrewed all their length, commonly called *wood ſcrews*. The central or working part of of the *gudgeon* ſhould always be 1 inch deeper than a ſemicircle, and project about 4 inches beyond the ſides of the *beam*; for particulars ſee page 71. The chair for the *gudgeon* will be eaſily underſtood by referring to the plan plate 5, fig. 20. 21.

GRATE BARS and BEARING BARS, plate 5, *fig.* 11, 12.

The grate bars to be caſt 6 inches deep, 3 inches broad at top and 1½ at the bottom, and ſhould have *knobs* at one end to hold them an inch or 1¼ inch aſunder, and *knobs* alſo in the middle of them 3 inches deep, to prevent them from bending.

The *bearers* of the *grate bars* ſhould be about 6 inches by 4; for the length of all ſee page 75.

HOTWELLS, plate 5, *fig.* 8, 9.

For the common engine to be caſt ¾ of an inch thick, and the height of all 3 feet, the outſide length of all at the top 4 feet, and at the bottom 18 inches, and the *ſpout* to project 4 inches. The center of the feeding pipe on the ſides of the hotwell to be all fixed 13½ inches high, and to project about 9 inches; they muſt not ſtand at *right angles* to the hotwell, but muſt ſlope inwards, forming an angle of 40 degrees, and ſhould incline downwards about 20 degrees, to point to the boiler; for particulars ſee page 71.

For the valve engine plate 4 and fig. 7, *(c)* ſhews their conſtruction.

INJECTION PIPES, plate 5, *fig*. 25.

Shews one of them with a branch upon it, to feed the piſton, and ſhould be made as much as poſſible of caſt iron, the metal $\frac{5}{8}$ of an inch thick, and that of the *flanges* $\frac{7}{8}$ thick. The *injection pipe* which lies within the cylinder of the common engine, ſhould be alſo of caſt iron, metal $\frac{1}{2}$ an inch thick, and wedged in the pipe caſt on the cylinder bottom, and the diameter of theſe pipes may be 2 inches, leſs than the directions for injection pipes given page 72. One end of this *ſmall pipe* is caſt cloſe, and a ſmall door is fixed on the upper ſide of it, as deſcribed plate 5, fig. 15, This door is covered with a plate of malleable iron about $\frac{3}{8}$ of an inch thick, and in the middle of this plate is cut a *ſquare hole* for the purpoſe of injecting, which plate may be adjuſted by raiſing the ſide with leather to make the *jet* ſtrike fair on the center of the piſton; the ſize of the *injecting hole* is given page 72.

For the valve engine the method of injecting is ſhewn plate 4, fig. 7.

JACK

JACK HEAD WORKING BARRELS.

Performing $\frac{2}{3}$ of the length of the full *ſtroke* of the engine, the diameters of them are given page 72, which will be found to afford water enough for condenſing. The length of the *working barrel* for a 6 feet ſtroke ſhould be 8 feet, and metal $\frac{7}{8}$ thick when bored, and *flanges* 1$\frac{1}{8}$ inch thick. The ſtrength of the *jack head ſmooth rods* are given page 72.

INJECTION COCKS.

To be made of braſs with *ſquare ſhanks*; they are uſed for the common engine only; ſee the water way of them page 72.

JACK HEAD PUMPS.

The metal of them may be $\frac{5}{8}$ or $\frac{3}{4}$ thick, and the joints may be either ſpigot and faucet, or hoboy joints run with *lead* and *regulus*. The diameter of theſe do not require to be more than the working barrel, as no rod works in them.

MAIN CHAINS.

To fix to the martingals of the *regulator beam*, ſee the plan of a link plate 4, fig. 4, on an enlarged ſcale.

Length of each chain muft be 9 feet 9 inches, the links *three* and *two*, and meafures 6½ inches long from center to center of the pin; for the ftrength of them fee page 72, and obferve that the annexed weight is given for one end of the *beam*.

MARTINGALS.

To fix to the regulator beam and main chains; two are required at each end, and fhould be about 5 feet 6 inches long, and the annexed weight is given for one end of the *beam*, fee page 73.

MAN-HOLES.

To make a road into the boiler, fhould be a *plain pipe* 2 feet 6 inches long and 21 inches diameter, and the metal ¾ thick, with a *flange* at each end 3 inches broad and 1 inch thick, containing 8 holes 1 inch fquare.

PUMPS, PIPES, or TREES.

The plain pipes to be caft 9 feet long each, and the ftrength may be proportioned as follows. The firft 4 pipes fhould be ¾ of an inch thick on the fide, and the *flanges* 1 inch thick and 3 inches broad. The fecond 4 pipes may be ⅞ thick and the *flanges* as above. The next

4

4 pipes may be 1 inch thick, and the *flanges* 3 inches broad and 1$\frac{1}{8}$ thick. The next two or three pipes which extend down as low as 42 or 45 yards, (and are as long as the common pipes of any ſet ought to be,) may be 1$\frac{1}{8}$ inch thick, with *flanges* 3 inches broad and 1$\frac{1}{4}$ thick. The weight, being put upon the pumps, is ſufficient to diſtinguiſh them, and the holes in the *flanges* for 18 fathoms down, ſhould be 1$\frac{1}{4}$ inch ſquare to take *bolts* 1$\frac{1}{8}$ ſquare. A pump 10 inches diameter ſhould have 6 holes in the *flange*, a 13 inch pump ſhould have 8 holes, and a 16 inch pump may do with 8 holes alſo.

The BUCKET and CLACK TREES, ſee plate 5, *fig.* 29, 30, 31, 32.

Should be caſt 6 feet long each, and ſuppoſing the whole ſet of pumps to be 50 or 54 yards deep, the metal ſhould be as follows. The plain part of the pipe ſhould be 1$\frac{1}{4}$ thick, the ſwelled part at the door 1$\frac{1}{2}$ thick, the projecting part of the door 1$\frac{3}{4}$ thick, the *flange* of the door 2$\frac{1}{2}$ thick, and the *flanges* of the pipe 1$\frac{1}{2}$ inch thick, and the bolt (which muſt be cloſe to the projecting part) ſhould be 2 inches ſquare, to take bolts 1$\frac{7}{8}$ ſquare.—The door front or face ſhould project 2$\frac{1}{2}$ inches from the ſide of the pump, and ſhould have a bead projecting out

out 1 inch at the bottom for the door to reſt upon. The *clear height* of the door may be 19 inches, and ſhould be 1 inch wider than the working barrel. The ſtrength of the malleable iron croſs bars for fixing on the door, are alſo given page 74, ſuitable to a ſet of pumps 50 or 54 yards deep and may be a little diminiſhed for ſhorter ſets.

The WORKING BARRELS, plate 5, *fig.* 28.

Should be caſt 11 feet long and the metal left 1¼ thick when bored, and the *flanges* 1½ thick; and the top end ſhould be *bell mouthed* ½ an inch on each ſide, to prevent the buckets from catching.

The WIND BORES, plate 5, *fig.* 35.

May be caſt 8 feet long with a plain or egg bottom as occaſion requires; the metal 1⅜ thick and the *flanges* 1⅜ alſo. The ſwelled part may be 3 feet long, with holes in it about 2 inches diameter, exceeding the *area* of the pipe.

DIRECTIONS for FIXING the DIAMETER of PUMPS.

The *common pipes* ſhould be 1 inch larger than the working barrel.

The

The *bucket pipes* the ſame as the common pipes.

The *clack pipe* below the ſeat, to be 1 inch leſs than the working barrel, and the ſeat at the top to be the ſame ſize with the working barrel, which ſhould taper $\frac{1}{8}$ of an inch on each ſide to an inch deep, and the depth 4 inches.

The *wind bore pipe* ſhould be 1 inch leſs than the working barrel.

PILLARS under the CYLINDER, ſee plate 5, fig. 14.

For the *common engine* are 5 feet 3 inches long, and for the *valve engine* only 2 feet 5 inches long; to have a hole through them $1\frac{1}{2}$ inch diameter, and the thickneſs of the metal as given page 73.

PLUG TREES.

To work the regulator and injection of the engine, take them about 21 feet long, and ſcantling $9\frac{1}{2}$ inches by $4\frac{1}{2}$.

PISTONS. ſee plate 5, fig. 16, 17.

Should be $\frac{1}{8}$ or $\frac{3}{16}$ leſs than the cylinder, the *ſtuffing ring* ſtands 4 inches from the ſide, and ſhould be 1 inch thick at the bottom, and $\frac{3}{8}$ or $\frac{1}{2}$ inch at the top. There ſhould be 4 ſquare holes for the ſhanks at right angles to each other,

other, at 3 inches from the *rim*, or *one ſtrong hole* in the center. A 60 inch cylinder requiring 6 piſton weights, each 3 inches ſquare, ſhould have 12 holes 1 inch ſquare cloſe within the *rim*, to put in *bolts* for ſcrewing down the *weights*, and 1 hole more to let out the water occaſionally; and the bottom of the piſton ſhould be caſt *a little convex* to diſperſe the water: for other particulars of the piſtons and weights ſee pages 69 and 72.

PISTON SHANKS.

Suppoſing 4 to each piſton, ſhould ſtand about 3 feet high, when fixed: the ſtrength of them is given page 73.

REGULATOR BEAMS, plate 4, fig. 1, 3.

The length from center to center of the chains to be 25 feet, and I would adviſe, where it may be had, to have them in one piece of *oak*. The ſides of the *beam* ſhould be a little rounded, as alſo the top of it, in the length way; and in the breadth way of it, the top ſide may be rounded, which leaves the wood in the ſtrongeſt ſtate poſſible; for the *ſcantling* of the beams ſee page 72, and to fix a beam curved as deſcribed plate 4, the center of the *gudgeon* requires to be raiſed 30 or 31 inches above the

the frame laid in the wall, to give a 9 feet *ſtroke*, and allow proper height for the *inſide ſpring beams* (24 inches thick) and about 6 inches for the ſprings.

The HEADS for the BEAM.

Are ſhewn plate 4, and require to be 10½ feet long; the ſcantlings are given page 72.

RECEIVERS, ſee plate 5, *fig.* 4, 7.

For the common engine require the height of all, on the ſides, to be 20 inches, the *ſteam pipe* above the *receiver* to ſtand up 9 inches, including the *flange*; and the diameter of the ſteam pipe to be the ſame with thoſe placed upon the cylinder bottoms herein before explained, page 48. The projecting pipes that point towards the boilers, ſtand out 20 inches, and droop 1½ inch at the end; and their diameters are the ſame as the communicating pipes before deſcribed. The *flanges* of all receivers ſhould be 3½ inches broad; their thickneſs is given page 73, and the weight of them given in that table includes the regulator plate and the receiver bottom.

REGULATOR PLATES, plate 4, *fig.* 7.

Theſe plates form the top of the *receivers*, the cock hole of which, in all regulator plates, muſt ſtand 3 inches to the right of the center of the communicating pipe, (looking towards the boiler,) and muſt be 6 inches deep, 1½ of which ſtands below, and 3½ above the plate. The face of the ſteam pipe muſt alſo project 1½ inches downwards, the cock hole muſt have ⅛ of an inch taper in every inch deep; and there muſt be a ſmall piece of the low *flange* of the cylinder bottom and the neck of the receiver taken out, to give room for the cock to work. The Founder will be careful to place the hole diſtinguiſhed by *, *fair* over the *middle* of the communicating pipe, and the firſt hole marked with † muſt be *fair* in a line with the hole marked * in the *receiver flange*, to make the *receiver ſteam pipes* ſtand ſquare with the engine houſe.

The diameters of all receivers are given page 73, with their thickneſs of metal, &c. and the poſition and diameter of the cock holes in the regulators, are given in the ſame page. The thickneſs of all regulator plates may be 1 inch, and the regulators ſhould be fixed with *a ſpring* to keep them as tight as poſſible.

SINK PIPES, ſee plate 5, fig. 6.

For the common engines, the metal of them may be ¾ thick, their diameters are given page 73, and the *ſink pipe* of the *valve engine* is diſtinguiſhed, plate 4, fig. 7, by dotted lines.

SPEARS with PLATES and RODS.

The *ſplicing* of the joints ſhould be 4 feet long; and every other particular for the *ſpears* &c. is explained page 74, and obſerve that the ſtrength of the U plates muſt be the ſame as the *ſpear plates.*

WASTE WATER PIPES.

The *top pipe* which joins to the *jack head ciſtern,* is ſhewn plate 5, fig. 26; the metal of all may be $\frac{5}{8}$ of an inch thick; and the ſize of them is given page 73.

The WOKRING GEER.

A front and ſide view for the common engine are given plate 4, fig. 2. The height of the *regalator axis* is 3 feet, that of the firſt *injection iron axis* 4 feet 3 inches, of the ſecond do. 6 feet 6 inches, and that of the third axis is 7 feet 5 inches. The width of the *plug frame* within the poſts is 3 feet, and from the center of the poſts to the center of the *moving plug,* 14 inches.

SPECIMEN of an ORDER to the BRASS FOUNDER.

For the Articles of a 60 inch Cylinder.

1 Injection cock with ſquare ſhanks, 4$\frac{1}{4}$ inches by 1$\frac{1}{4}$ inch water way.
2 Feeding cocks made in the ſtop cock way, with a hole in the key to turn by hand occaſionally. 2 Steam vales 4 inches the leaſt diameter.
1 Snifting cock 2 inches, inſide diameter made the ſtop cock way.
1 Piſton cock (bib) 2 inches inſide diameter. 2 *Gauge* cocks (bib) 1$\frac{1}{4}$ do.
1 Air cock (bib) to fix in the ſink pipe $\frac{3}{4}$ inch inſide diameter.
1 Jack head vale 5 inches the leaſt diameter.

The table given in the pages 70 to 73 inclufive, fhews the length and ftrength &c. of all the materials of an engine of every fize of cylinder, rifing 5 inches in the diameter at a time. For example, a 60 inch cylinder requires 2 boilers each $14\frac{1}{2}$ feet diameter, with 2 crofs beams to fcrew down the cylinder each 11 feet 4 inches long, and fcantling 15 inches by 10, (but the valve engine requires 3 of them,) 2 cylinder beams 10 feet 6 inches long and 20 by 18, and fo on for every other article.

DIRECTIONS

FOR

BUILDING ENGINE HOUSES.

AFTER Digging the foundation of the houfe, the firft thing wanted of the carpenter, is a door cafe for taking out the afhes, which fhould be 3 feet wide; the centers ufed by the bricklayer for the openings in the fide walls, that give the communication to the fires, muft be for boilers of every fize, 3 feet 8 inches wide in the narroweft part, a platform of which is defcribed in the plan, plate 4, fig. 8; the height to the top of the arch (as 6 inches are fuppofed to be buried in the pitching) fhould be 6 feet

feet 9 inches or 7 feet. A hole to receive the coals muſt be left cloſe to the aſh hole door caſe; the ſteps from the aſh place to the *plug floor* may be taken up with the ſide walls of the houſe, which may be about 20 inches long; and a ſmall window ſhould be put in, under the plug floor to give light to the fire man. The upper ſides of the *croſs beams* which are intended to ſcrew down the cylinder of the common engine, muſt lie exactly 1 foot 7 inches below the upper ſide of the *plug floor;* but in the *valve engine* the *croſs beams* form a part of the floor. The height from the foundation of the houſe to the upper ſide of the plug floor, in both kinds of engines, is explained page 44. In laying the *plug floor* of the common engine (which muſt be pretty ſtrong where the hotwell ſtands) the ſpace muſt be left open, under the receiver, which is formed by the two croſs beams, for a road into the *receiver.* In the common engines the ſpaces in the ſide walls of the houſe for the communicating pipes to lay through, muſt be ſet out at right angles from the center of the cylinder, the width of 15, 16, or 18 inches, and ſhould begin at 9 inches below the *plug floor,* and be taken up as high as 2 feet 6 inches above it, which ſpaces ſhould be left 5 inches, on each ſide, wider in the inſide of the houſe, for a brick in length, to give room to wedge the ring tight; but for the valve engine, the communicating pipes go off

at

at right angles, from the center of the *ſteam cheſt*, and the ſpaces in the ſide walls ſhould begin at 12 inches above the *plug floor* and be taken up 3 feet 6 inches high: as in this engine the feeding pipes cannot be taken through the ſpaces left for the communicating pipes, a hole muſt be left 8 inches broad, beginning 4 inches above the floor, muſt be taken about 16 inches high, and ſhould be ſet off at right angles, from 21 inches beyond the center of the cylinder. At the level of the *plug floor* muſt be left a cavity in the end wall of the houſe for taking in the cylinder, and it would be well to lay in a plank, at the level of the *plug floor*, to preſerve the brick work, and this ſpace or opening muſt be carried up as high as the frame that lies round the building, at the height of the under ſide of the cylinder beams deſcribed plate 4, fig. 5, by ſo much of it as is drawn by parallel lines; the end of this frame (which ſhould be from 10 to 14 inches ſquare) will ſerve as a *foundation* or *lentil*, to build the end wall upon. The door into the *plug floor chamber* muſt not be forgotten, nor a window at 5 feet above the *plug floor*, to give light to the plug man, and after the cylinder and piſton are got up, it would do well in the walling up of that cavity, to put in another window there, leaving a ſmall ſpace for the convenience of taking away the hot water.

The

The upper ſides of the cylinder beams muſt lie 11 feet 7 inches high, above the plug floor, in the common engines, and thoſe of the valve engines only 9 feet 8 inches high. Spaces muſt be left in the ſide walls of the houſe to admit theſe cylinder *beams*, which ſpaces muſt be left wide enough for the *cup* of the cylinder to paſs through, and ſo narrow as not to prevent the wedging of theſe beams to the ſides of the cylinder; and muſt be taken up 4 or 6 inches above the cylinder floor, to enable the carpenter to wedge down the *cylinder beams*. Holes muſt alſo be left in the walls at the height of the cylinder beams, to receive the joiſts of the *cylinder floor*, and a window muſt be fixed in the end wall of the building, to light that chamber: at the height of 9 feet 6 inches above the cylinder floor, (provided the *regulator beam* is ſomewhat curved like the beam deſcribed in the plan, plate 4, fig. 1,) muſt be laid in another frame of wood, the ſame ſize as the lower one, to ſtrengthen the building, but with the addition of the dotted lines to it, deſcribed plate 4, fig. 5, meaſuring to the upper ſide of the frame; and if the *beam* ſhould be ſtraighter, the frame muſt be laid a little lower. Upon this frame are laid the double *ſpring beams* for the inſide of the houſe, and 2 of them go 18 or 20 inches through the main wall for the convenience of fixing the outſide ſpring frame. At the level of the upper ſide of the *lower*

ſpring

ſpring beams, is laid the *beam floor,* and apertures muſt be left on the ſide walls for the joiſts, and at this frame the walls of the houſe are reduced half a brick on each ſide. The ſide walls being raiſed up, about 20 feet above the *cylinder floor,* will give ſufficient height for the working of the *regulator beam,* obſerving alſo that another window muſt be put in this chamber, and a cavity left in the main wall for the regulator beam, with a ſpace for a road on each ſide of it, which cavity ſhould be taken 8 or 8½ feet high above the laſt mentioned frame, to the top of the arch. In roofing the houſe a coupling ſhould be fixed directly over the center of the cylinder, which will alſo accommodate the *piſton blocks,* and a projecting piece of wood muſt be laid to the beam of the coupling, to fix or *ſteady* the *boiler chimney.* It is unneceſſary to mention that 4 or 6 projecting pieces of wood may be laid in the front wall to *fix* and *ſteady* the *ſhear legs,* and 6 or 8 more pieces to fix the jack head pipes. The jack head or ciſtern pillar muſt be raiſed upon the main wall, and 2 pieces of wood laid acroſs it, at 4 or 5 feet above the roof of the houſe, for the convenience of fixing down the *ciſtern* by bolts; the top of this ciſtern ſhould be raiſed 46½ feet above the *plug floor*; and the bricklayer is deſired to run all the walls of the building as high as the beam floor, with putty, to unite them well together.

SEE PAGE 59.

Diameter.	2½ *In.* thick middle, 1 sides.			*Diameter.*	3 *In.* thick middle, 1¼ sides.			*Diameter.*	3½ thick middle, 1½ sides.			*Diameter.*	4 *In.* thick middle, 1¾ sides.			*Diameter.*	4½ thick middle, 2 sides.			*Diameter.*	5 *In.* thick middle, 2 sides.		
In.	*c.*	*q*	*lb.*	*In.*	*c.*	*q*	*lb.*	*In.*	*c.*	*q.*	*lb.*	*In.*	*e.*	*q.*	*lb.*	*In.*	*c.*	*q*	*lb.*	*In.*	*c.*	*q*	*lb.*
15		3	12	25	2	2	17	35	6	0	8	45	11	1	2	55	18	2	25	65	27	3	16
16		3	26	26	2	3	20	36	6	1	21	46	11	3	2	56	19	1	18	66	28	3	4
17	1	0	12	27	3	0	23	37	6	3	6	47	12	1	3	57	20	0	12	67	29	2	20
18	1	0	26	28	3	1	25	38	7	0	19	48	12	3	5	58	20	3	6	68	30	2	8
19	1	1	14	29	3	3	0	39	7	2	5	49	13	1	9	59	21	2	2	69	31	1	25
20	1	2	2	30	3	3	21	40	7	3	19	50	13	3	13	60	22	0	26	70	32	1	15
21	1	2	20	31	4	0	23	41	8	1	8	51	14	1	18	61	22	3	24	71	33	1	6
22	1	3	10	32	4	1	26	42	8	2	25	52	14	3	25	62	23	2	23	72	34	0	21
23	1	3	27	33	4	3	1	43	9	0	15	53	15	2	5	63	24	1	23	73	35	0	18
24	2	0	16	34	5	0	4	44	9	2	5	54	16	0	13	64	25	0	24	74	36	0	11
—	—	0	—	—	—	0	—	—	—	0	—	—	—	0	—	—	—	0	—	75	37	0	6

TABLE shewing the WEIGHT and THICKNESS of CYLINDER BOTTOMS,

(The depth of all for the Common Engine 12 inches, including the flange, and the Lugs 6 inches square)

SEE PAGE 48.

Sides ⅞ thick, Bottom 1⅛ do. Flange 1⅛ do. and 3½ broad, Lugs 1¼ thick.				Sides ⅞ thick Bottom 1⅛ do. Flange 1⅛ do. and 3½ broad, Lugs 1½ thick.				Sides ⅞ thick. Bottom 1¼ do Flange 1¼ do and 4 broad, Lugs 1¾ thick.				Sides 1 In. thick. Bottom 1½ do. Flange 1½ do. and 4 broad, Lugs 2 thick.				Sides 1⅛ thick Bottom 1¾ do. Flange 1¾ do and 4 broad, Lugs 2½ thick.				Sides 1⅛ thick. Bottom 1¾ do. Flange 1¾ do. and 4 broad, Lugs 2½ thick.			
Di.	*c.*	*q.*	*lb.*	*Di.*	*c.*	*q.*	*lb.*	*Di.*	*c.*	*q.*	*lb.*	*Di.*	*c.*	*q.*	*lb.*	*Di*	*c.*	*q*	*lb.*	*Di.*	*c.*	*q*	*lb.*
15	4	1	—	26	6	2	18	36	9	2	4	46	15	1	—	56	22	0	19	66	27	2	5
16	4	1	26	27	6	3	16	37	9	3	6	47	15	2	15	57	22	2	20	67	28	0	13
17	4	2	24	28	7	0	15	38	10	0	6	48	16	0	2	58	23	0	21	68	28	2	22
18	4	3	22	29	7	1	11	39	10	1	11	49	16	1	18	59	23	2	23	69	29	1	3
19	5	0	20	30	7	2	8	40	10	3	4	50	16	3	7	60	24	0	26	70	29	3	14
20	5	1	19	31	7	3	6	41	11	0	7	51	17	0	25	61	24	3	1	71	30	1	24
21	5	2	18	32	8	0	3	42	11	1	13	52	17	2	14	62	25	1	5	72	31	0	7
22	5	3	17	33	8	1	0	43	11	2	19	53	18	0	5	63	25	3	11	73	31	2	19
23	6	0	16	34	8	1	26	44	11	3	26	54	18	1	23	64	26	1	16	74	32	1	4
24	6	1	15	35	8	2	23	45	12	1	4	55	18	3	16	65	26	3	23	75	32	3	17
25	6	2	14	—	—	0	—	—	—	0	—	—	—	0	—	—	—	0	—	—	—	0	—

SEE PAGE 64.

Diameter of Cylinders.	BOILERS. *page* 43.		CROSS BEAMS Under Pillars. *page* 50.			CYLINDER BEAMS. *page* 51.				CISTERNS *for the* JACK HEAD. *page* 51.						
	Number.	Diameter.	Length.		Scantling.	Length.		Scantling.	Quantity.	Length.		Breadth.		Height.		Thickness of Planks.
In.		*Feet.*	*Feet*	*In.*	*In. by In.*	*F.*	*In.*	*In. by In.*	*F.*	*F.*	*In*	*F*	*In*	*F*	*In.*	*In.*
25	1	8½	9	6	9 by 9	8	3	13 by 9	14	5	6	3	9	4	6	2
30	1	10½	10	0	10 by 10	8	3	14 by 10	16½	6	0	3	9	4	9	2
35	1	12½	10	0	10 by 10	8	6	15 by 11½	20	6	3	4	0	4	9	2
40	1	14	10	6	11 by 10	8	9	16 by 13	25	6	6	4	0	4	9	2¼
45	2	11	9	6	11 by 10	9	1	17 by 14½	32	6	6	4	3	4	9	2¼
50	2	12½	10	0	12 by 10	9	6	18 by 16	38	6	6	4	3	4	9	2½
55	2	13½	10	6	13 by 10	10	0	19 by 17	45	6	6	4	6	5	0	2½
60	2	14½	11	0	14 by 10	10	6	20 by 18	52	6	6	4	6	5	6	2½
65	2	16	11	4	15 by 10	11	0	21 by 19	61	6	9	4	9	5	9	2½
70	2	17	11	8	16 by 11	11	6	22 by 20	68	7	0	5	0	6	0	2½

Diameter of Cylinders.	ENGINE HOUSE DIMENSIONS. *page* 52. INSIDE.				STRENGTH *of* WALLS.			No. of Bricks. *page* 52.	Lime. *page* 52.	Sand. *page* 52.	Masons Bills. *page* 52.	Feeding Pipes for the BOILERS.	
	Length.		Breadth.		Front.	Sides.	End.					Number.	Inside Diam.
In	*F.*	*In.*	*F.*	*In.*	*In. or In.*	*In.*	*In.*	*Thods.*	*Ch.*	*Loads*	*£.*		*In*
25	15	6	7	0	29 or 30	19	15	48	16	32	20	1	2½
30	16	0	7	0	34 or 35	19	19	55	18	36	22	1	3
35	16	0	7	4	38 or 39	24½	19	65	21	42	24	1	3
40	16	0	7	6	38 or 39	24½	19	70	23	46	26	1	3½
45	16	4	8	0	43 or 44	24½	24½	80	27	54	32	2	3
50	16	8	8	4	48 or 49	24½	24½	90	30	60	39	2	3
55	17	0	8	10	53 or 54	29	24½	102	36	72	45	2	3¼
60	17	6	9	3	58 or 60	29	24½	120	40	80	55	2	3½
65	17	10	9	8	68 or 70	34	29	134	46	92	66	2	3¾
70	18	3	10	1	77 or 78	34	29	150	54	108	75	2	4

SEE PAGE 64.

Diameter of Cylinders	COMMUNICATING PIPES. *page* 51.							CATCH PINS. *page* 47.				Scantling of frame in the walls that the Cylinder beams rest on. *page* 52.	Height these frames lie in the common engine.	
	Number.	Inside Diam.	Length.		Weight.			Length.		Scantling.	Weight.			
In.		*Inch.*	*Feet*	*In.*	*c.*	*q.*	*lb*	*F.*	*In.*	*In.* by *In.*	*lbs.*	*In* by *In*	*F*	*In.*
25	1	8	6	2	3	3	20	2	7	2¾ by 1¾	86	6 by 3	10	5½
30	1	9	6	10	4	3	—	2	8	3 by 2	107	6 by 3	10	4½
35	1	10	8	2	6	0	21	2	9	3¼ by 2	120	6 by 3	10	3½
40	1	12	8	11	8	0	0	2	10	3½ by 2⅛	142	9 by 3	10	2½
45	2	9	7	10	10	1	4	3	0	3¾ by 2⅛	161	9 by 3	10	1½
50	2	10	8	5	12	3	4	3	2	4 by 2¼	190	9 by 3½	10	0½
55	2	11	9	3	15	1	8	3	3	4¼ by 2¼	210	9 by 3½	9	11½
60	2	12	9	9	17	3	4	3	4	4¾ by 2¼	240	9 by 6	9	10½
65	2	12½	10	11	19	0	6	3	5	5 by 2½	287	9 by 6	9	9½
70	2	13	10	—	19	3	8	3	6	5½ by 2½	323	9 by 6	9	8½

Diameter of Cylinders	GUDGEONS FOR BEAM CENTER. *page* 52.										HOTWELLS. *page* 53.						
	Diameter of working part	Length.		Breadth.		Thickness of met. in middle	Thickness of met. at ends.	Weight.			Outside Breadth.		Width of the Spout.	Depth of the Spout.	Weight.		
In	*In.*	*F.*	*In.*	*F*	*In.*	*In.*	*In.*	*c.*	*q*	*lb.*	*F.*	*In.*	*In*	*In.*	*c.*	*q*	*lb.*
25	3¾	3	4	1	9	1½	1	3	0	4	1	9	12	6	6	3	7
30	4	3	6	1	10	1¾	1	3	2	14	1	9	13	6	6	3	7
35	4¼	3	6	1	11	1¾	1	3	3	4	1	10	14	6	7	0	4
40	4½	4	0	2	1	2	1¼	5	0	20	1	10	14	6	7	0	4
45	4¾	4	0	2	3	2	1¼	5	3	21	1	11	15	7	7	1	18
50	5	4	3	2	6	2¼	1¼	7	1	23	1	11	15	7	7	1	18
55	5½	4	3	2	9	2¼	1½	8	3	17	2	0	16	8	7	2	10
60	6	4	6	3	0	2½	1½	10	3	27	2	0	16	8	7	2	10
65	6¼	4	6	3	1	2½	1½	11	2	—	2	2	17	8½	7	3	14
70	6½	4	8	3	2	2¾	1½	12	3	12	2	2	17	8½	7	3	14

SEE PAGE 64.

Diameter of Cylinders.	INJECTION PIPES. *page* 54. Diam. of common Injection Pipes.	Dia. of Injec. Hole fixed on little Pipe.	Length of the little Pipe within the Cylinder.		Diam. of Pipe on Cylin. bottom. 49	BRASS INJECTION COCKS. WATER WAY. *page* 55		Jack Head Working Barrels. *p.* 55. Diam.	MAIN CHAINS. *page* 55. All of them 3 and 2. No. of Rowes.	Scantling of the Three's.		Scantling of the Two's.		Dia. of Pins	Weight of one Chain.
In.	*In.*	*In.*	*F.*	*In.*	*In.*	*In.*	*In.*	*In.*		*In.*	*In.*	*In.*	*In.*	*In.*	*lbs.*
25	3 1/4	3/4	1	2	4 1/2	2 3/4 by	7/8	3 3/4	1	1 1/8 by	1/2	1 1/8 by	3/4	1 1/4	232
30	3 1/2	7/8	1	4	4 1/2	3 by	1	4 1/2	1	1 1/4 by	5/8	1 1/4 by	15/10	1 3/8	315
35	3 3/4	1	1	6	5	3 1/8 by	1	5 1/4	1	1 3/8 by	5/8	1 3/8 by	15/16	1 1/2	360
40	4	1 1/8	1	8	5	3 1/4 by	1	6	1	1 1/2 by	3/4	1 1/2 by	1 1/8	1 3/4	468
45	4 1/4	1 1/4	1	10 1/2	5 1/4	3 3/8 by	1 1/8	6 3/4	1	1 5/8 by	7/8	1 5/8 by	1 5/16	1 7/8	590
50	4 1/2	1 3/8	2	1	5 1/4	3 1/2 by	1 1/8	7 3/4	1	1 3/4 by	1	1 3/4 by	1 1/2	2	720
55	4 3/4	1 1/2	2	3 1/2	5 1/2	3 7/8 by	1 1/8	8 1/4	2	1 1/2 by	3/4	1 1/2 by	1 1/8	1 3/4	936
60	5	1 5/8	2	6	5 1/2	4 1/4 by	1 1/4	9	2	1 1/2 by	7/8	1 1/2 by	1 5/16	1 7/8	1114
65	5 1/4	1 3/4	2	8	5 3/4	4 1/2 by	1 1/4	9 3/4	2	1 5/8 by	1	1 5/8 by	1 1/2	1 7/8	1320
70	5 1/2	1 7/8	2	10	5 3/4	4 3/4 by	1 3/8	10 1/2	2	1 3/4 by	1	1 3/4 by	1 1/2	2	1440

Diam. of Cylin.	PISTON WEIGHTS, *with Handles. p.* 59, Scantling.	Weight.			REGULATOR BEAMS, *Oak,* 25 *feet long. p.* 60. Scantling in the Middle.		Scantling at the Ends.		Quantity.	BEAM HEADS, *Oak,* 10 1/2 *ft. long.* 61 Breadth.	Depth.	Quantity.	Jack Head smooth Rods *page* 55. Diam.
In.	*In.*	*c.*	*q.*	*lb.*	*In.*	*In.*	*In.*	*In.*	*Feet.*	*In.*	*In.*	*Feet.*	*In.*
25	3	1	1	20	22 by	20	19 by	15	70	15	10	23	1 7/16
30	3	1	2	—	24 —	22	21 —	16	80	16	11	27	1 1/2
35	3	2	0	7	26 —	23	22 —	17	95	17	12	31	1 9/16
40	3	2	1	—	28 —	25	24 —	18	110	18	12 1/2	35	1 5/8
45	3	2	2	24	30 —	27	26 —	19	125	19	13	38	1 11/16
50	3	2	3	—	32 —	29	28 —	21	140	21	14	44	1 3/4
55	3	3	1	13	33 —	30	29 —	22	155	22	14 1/2	48	1 7/8
60	3	3	2	—	35 —	32	31 —	24	175	24	15	54	2
65	3	4	0	2	36 —	33	32 —	25	190	25	15 1/2	60	2 1/8
70	3	4	2	—	38 —	35	33 —	26	210	26	16	65	2 1/4

FIRE ENGINE MATERIALS PROPORTIONED.

SEE PAGE 64.

Diameter of Cylinders.	FOUR MARTINGALS. p. 56. Length of each 5 feet 6 inches.			FOUR PILLARS. 59. Under Cylinder.				PISTON SHANKS. page 60. Stands 3 feet high when fixed.					
	Scantling at one End.	Scantling at the other.	Weight	Thicknefs of Metal.	WEIGHT.			Length.		Scantling	WEIGHT.		
In.	In. by In.	In. by In.	lbs.	In.	c.	q.	lbs.	F.	In.	In.	c.	q.	lbs.
25	2¼ by 1	1¼ by ¾	59	¾	1	1	14	3	9	1¼	0	2	22
30	2½ by 1	1½ by ¾	67	¾	1	1	14	3	9	1½	1	0	2
35	2½ by 1¼	1½ by ¾	72	7/8	1	2	20	3	9	1¾	1	1	2
40	2½ by 1½	1½ by 1	97	7/8	1	2	20	3	10	2	1	3	10
45	2¾ by 1¾	1½ by 1	127	1	4	0	—	3	10	2¼	2	1	8
50	3 by 2	1½ by 1¼	144	1	4	0	—	3	11	2½	2	3	21
55	3½ by 2	1¾ by 1½	170	1⅛	4	2	20	3	11	2¾	3	2	6
60	4 by 2⅛	2 by 1½	212	1⅛	4	2	20	4	—	3	4	1	8
65	4½ by 2⅛	2¼ by 1½	240	1¼	5	1	16	4	—	3⅛	4	2	16
70	5 by 2⅛	2½ by 1¾	288	1¼	5	1	16	4	1	3¼	5	0	20

Diam. of Cylin.	RECEIVERS. page 61.							REGULATORS. Plates 1 In. thick. 61.			SINK PIPES. p. 62. Metal ¾ thick.				Wafte Water Pipe. page 63
	Infide Diam.		Thicknefs of Sides.	Thicknefs of Flanges	Diam. of St. Pipes.	WEIGHT		Diftance of Cock & fteam P. center	Top Diam. Cock hole.	Bottom Diameter of Cock hole.	Infide Dia.	WEIGHT.			Diam.
In.	F.	In.	In.	In.	In.	c.	q.	In.	In.	In.	In	c.	q.	lb.	In.
25	1	9	¾	1	8	8	3	7¾	3	2¼	5½	2	1	10	2½
30	1	10	¾	1	8	9	1	8	3¼	2½	6	2	2	8	2½
35	1	11	¾	1	8½	9	3	8¼	3¼	2½	6½	2	3	6	2½
40	2	0	¾	1	9¼	10	3	8½	3½	2¾	7¼	3	0	9	2¾
45	2	3	7/8	1¼	9¾	15	0	9¾	3½	2¾	7½	3	0	7	2¾
50	2	6	1	1¼	10½	18	1	11	3¾	3	8	3	1	2	3
55	2	9	1	1¼	11	20	2	12	3¾	3	8¼	3	1	13	3
60	3	0	1	1¼	12	22	2	13	4	3¼	8½	3	1	24	3¼
65	3	2	1	1¼	12½	25	1	13½	4	3¼	8¾	3	2	7	3¼
70	3	4	1⅛	1⅜	13	27	2	14	4¼	3½	9	3	3	2	3¼

Diameter of the Pumps in the Pit.	BUCKETS & CLACKS. 46, 47. *Tapers about 1/8 of an Inch to an In.*				BUCKET SHANKS. *page 47.*		BUCKET HOOPS. *page 47.*	
	Depth of the Shell.		Top thickness	Size of Shank Holes.	Strength in Common.	Strength at the Joints.	Depth of them.	Thickness at Bottom.
In.	*Mid.*	*Sides.*	*In.*	*In.* *In.*	*In* *In.*	*In.*	*In*	*In.*
6	3	3	3/8	2 3/4 by 3/8	2 by 1 1/4	2 Square	1	1/8
8	3 1/4	3 1/4	7/16	3 by 7/16	2 1/4 by 1 3/8	2 1/4	1 1/4	3/16
10	3 1/2	3 1/2	1/2	3 1/4 by 1/2	2 1/2 by 1 1/2	2 1/2	1 1/2	1/4
12	3 3/4	3 3/4	9/16	3 1/2 by 9/16	2 5/8 by 1 3/4	2 5/8	1 3/4	5/16
14	4	5	5/8	3 3/4 by 5/8	2 3/4 by 2	2 3/4	2	3/8
16	4 1/4	6	3/4	4 by 3/4	3 by 2 1/4	3	2 1/2	7/16
18	4 5/8	6 1/2	7/8	4 by 3/4	3 1/8 by 2 1/2	3 1/8	2 5/8	1/2
20	5	6 3/4	1	4 1/2 by 13/16 (2 Holes below.)	3 1/4 by 2 3/4	3 1/4	2 3/4	9/16
22	5 1/2	7 1/4	1 1/8	4 3/4 by 7/8	3 1/2 by 3	3 1/2	2 7/8	5/8
24	6	7 3/4	1 1/4	5 by 1	3 3/4 by 3 1/4	3 3/4	3	5/8

Diameter of the Pumps in the Pit.	SPEARS. *Fir. 63.*	SPEAR PLATES & BOLTS *page 63.*					Bucket Rods, *11½ f. long. 63.*		CROSS BARS & BOLTS, *for Bucket & Clack Doors. 57.*			
	Scantling.	Length.	Breadth.	Thickness in the middle.	Thickness at the ends.	Diameter of the Bolts.	*Square.* Scantling	Length of U. Plates.	Diameter of the Bolts.	Thickness of Bars in Middle	Thickness of Bars at Ends.	Breadth of the Bars.
In.	*In.*	*F.*	*In.*	*In.*	*In.*	*In.*	*In.*	*F.*	*In.*	*In.*	*In.*	*In.*
6	3 Square	6	2 1/2	3/8	3/16	5/8	1 1/2	4 1/2	1	3/4	1/2	2 1/2
8	3 1/2	6 1/2	2 3/4	7/16	3/16	11/16	1 3/4	4 3/4	1 1/8	1	3/4	2 1/2
10	4	7	3	1/2	1/4	3/4	2	5	1 3/8	1 1/4	7/8	2 1/2
12	4 1/2	7 1/2	3 1/2	9/16	5/16	3/4	2 1/4	5 1/4	1 1/2	1 3/4	1	3
14	5	8	3 1/2	5/8	3/8	7/8	2 1/2	5 1/2	1 3/4	2	1 1/4	3
16	5 1/2	8 1/2	4	3/4	1/2	7/8	2 3/4	6	1 7/8	2 1/2	1 3/4	3
18	6	9	4	13/16	9/16	15/16	3	6 1/2	2	3	2 1/4	3
20	6 1/2	9 1/2	4 1/2	7/8	5/8	1	3 1/4	7	2 1/4	3 3/8	2 1/2	3
22	7	10	4 1/2	15/16	11/16	1 1/8	3 1/2	7 1/2	2 3/8	3 3/4	2 3/4	3
24	7 1/2	10 1/2	4 3/4	1	3/4	1 1/4	3 3/4	8	2 1/2	4	3	3

Diameter of Boilers	ASH HOLES, *Under the Boilers. p. 45.*				TWO PLATES, *2 In. thick in middle, and 1½ at ends, to make up the space between door frame & bars. page 46.*									CHIMNEY PIPES. *p. 51*	GRATE BARS, *laid 2 f. 9 in. high. 53*			BEARING BARS. *page 53.*				
	Length.		Width.		Length of long side of the two Plates laid together.		Length of short side of the two plates laid together.		Breadth of the two plates when laid together.		Weight of the two Plates including Plate that lays over door frame.			Diam.	Number.	Length.		Number.	Length.		Weight of one Sett of each.	
Ft.	*Ft.*	*In.*	*Ft.*	*In.*	*Ft.*	*In.*	*Ft.*	*In.*	*Ft*	*In.*	*c.*	*q.*	*lbs.*	*In.*		*F.*	*In.*		*Ft*	*In.*	*c.*	*q.*
8	3	9	2	4	3	5	3	5	1	0½ (Cast in one Plate.)	3	1	—	16	7	3	9	3	3	4	16	2
8½	3	10½	2	6	3	7	3	7	1	2½	3	3	2	16	7	3	10½	3	3	6	17	1
9	4	0	2	8	3	9	3	9	1	3½	4	1	—	17	8	4	0	3	3	8	19	1
9½	4	1½	2	10	3	11	3	11	1	4½	4	3	—	17	8	4	1½	3	3	10	20	0
10	4	3	3	0	4	2	4	1	1	6	6	0	—	18	9	4	3	3	4	0	22	1
10½	4	4½	3	2	4	4	4	1	1	7	6	1	7	18	9	4	4½	3	4	2	23	0
11	4	6	3	4	4	6	4	1	1	8	6	2	16	19	10	4	6	3	4	4	25	1
11½	4	7½	3	6	4	8	4	1	1	9	6	3	26	19	10	4	7½	3	4	6	26	1
12	4	9	3	8	4	10	4	1	1	10	7	1	9	20	11	4	9	3	4	8	28	3
12½	4	10½	3	10	5	0	4	1	2	0	7	3	—	20	11	4	10½	3	4	10	29	3
13	5	0	4	0	5	2	4	1	2	1	8	1	—	21	12	5	0	3	5	0	32	1
13½	5	1½	4	2	5	4	4	1	2	2	8	2	—	21	12	5	1½	3	5	2	33	1
14	5	3	4	4	5	6	4	1	2	3	9	0	—	22	13	5	3	3	5	4	36	1
14½	5	4½	4	6	5	8	4	1	2	5	9	1	—	22	13	5	4½	3	5	6	37	0
15	5	6	4	8	5	10	4	1	2	6	9	3	—	23	14	5	6	3	5	8	40	0
15½	5	7½	4	10	6	0	4	1	2	7	10	1	—	23	14	5	7½	3	5	10	41	0
16	5	9	5	0	6	2	4	1	2	8	10	3	—	24	15	5	9	3	6	0	44	0
16½	5	10½	5	2	6	4	4	1	2	9	11	1	—	24	15	5	10½	3	6	2	45	0
17	6	0	5	4	6	6	4	1	2	10	11	3	—	25	16	6	0	3	6	4	48	0

SEE PAGE 42.

Pumps, Diameter in Inches.	Depth Shaft, Faths.	Diameter of the Cylinder.	Depth of Shaft in Fathoms.	Diameter of the Cylinder.	Gallons drawn at a 6 foot Stroke.	Strokes in 1 Minute.	Gallons drawn in One Minute.	WATER drawn in one Hour. Hogsheads	Gal.
7	5	9.6	55	31.7	9.83	7	68.8	65	33
	10	13.5	60	33.1		8	78.6	74	54
	15	16.6	65	34.5		9	88.4	84	12
	20	19.1	70	35.8		10	98.3	93	39
	25	21.3	75	37.0		11	108.1	102	60
	30	23.4	80	38.2		12	117.9	112	18
	35	25.3	85	39.4		13	127.7	121	39
	40	27.0	90	40.5		14	137.6	131	3
	45	28.7	95	41.6		15	147.4	140	27
	50	30.2	100	42.7		16	157.2	149	45
8	5	10.9	55	36.2	12.8	7	89.6	85	21
	10	15.4	60	37.8		8	102.4	97	33
	15	18.9	65	39.3		9	115.2	109	45
	20	21.8	70	40.8		10	128.0	121	57
	25	24.4	75	42.3		11	140.8	134	6
	30	26.7	80	43.6		12	153.6	146	18
	35	28.9	85	45.0		13	166.4	158	30
	40	30.9	90	46.3		14	179.2	170	42
	45	32.7	95	47.6		15	192.0	182	54
	50	34.5	100	48.8		16	204.8	195	3
9	5	12.3	55	40.7	16.2	7	113.4	108	—
	10	17.4	60	42.5		8	129.6	123	27
	15	21.3	65	44.3		9	145.8	138	54
	20	24.6	70	45.9		10	162.0	154	18
	25	27.5	75	47.5		11	178.2	169	45
	30	30.1	80	49.1		12	194.4	185	9
	35	32.5	85	50.6		13	210.6	200	36
	40	34.7	90	52.1		14	226.8	216	0
	45	36.8	95	53.5		15	243.0	231	27
	50	38.8	100	54.9		16	259.2	246	54

SEE PAGE 42.

Pumps, Diameter in Inches.	Depth Shaft, Faths.	Diameter of the Cylinder.	Depth of Shaft in Fathoms.	Diameter of the Cylinder.	Gallons drawn at a 6 foot Stroke.	Strokes in 1 Min.	Gallons drawn in One Minute.	WATER drawn in one Hour. Hogsheads	Gal.
	5	13.7	55	45.2		7	140.	133	21
	10	19.3	60	47.2		8	160.	152	24
	15	23.6	65	49.2		9	180.	171	27
	20	27.3	70	51.0		10	200.	190	30
	25	30.5	75	52.8		11	220.	209	33
10	30	33.4	80	54.5	20.	12	240.	228	36
	35	36.1	85	56.2		13	260.	247	39
	40	38.6	90	57.9		14	280.	266	42
	45	40.9	95	59.5		15	300.	285	45
	50	43.1	100	60.1		16	320.	304	48
	5	15.0	55	49.8		7	169.4	161	21
	10	21.2	60	52.0		8	193.6	184	24
	15	26.0	65	54.1		9	217.8	207	27
	20	30.0	70	56.1		10	242.0	230	30
	25	33.5	75	58.1		11	266.2	253	33
11	30	36.8	80	60.0	24.2	12	290.4	276	36
	35	39.7	85	61.9		13	314.6	299	39
	40	42.4	90	63.7		14	338.8	322	42
	45	45.0	95	65.4		15	363.0	345	45
	50	47.4	100	67.1		16	387.2	368	48
	5	16.4	55	54.3		7	201.6	192	—
	10	23.1	60	56.7		8	230.4	219	27
	15	28.3	65	59.0		9	259.2	246	54
	20	32.7	70	61.2		10	288.0	274	18
	25	36.6	75	63.4		11	316.8	301	45
12	30	40.1	80	65.5	28.8	12	345.6	329	9
	35	43.3	85	67.5		13	374.4	356	36
	40	46.3	90	69.4		14	403.2	384	—
	45	49.1	95	71.3		15	432.0	411	27
	50	51.8	100	73.2		16	460.8	438	54

 CYLINDERS *proportioned to sundry Depths, and Sizes of* PUMPS.

SEE PAGE 42.

Pumps, Diameter in Inches.	Depth Shaft, Faths.	Diameter of the Cylinder.	Depth of Shaft in Fathoms.	Diameter of the Cylinder.	Gallons drawn at a 6 foot Stroke.	Strokes in 1 Min.	Gallons drawn in One Minute.	WATER drawn in one Hour. Hogsheads	Gal.
	5	17.7	55	58.9		7	236.6	225	21
	10	25.0	60	61.4		8	270.4	257	33
	15	30.7	65	63.9		9	304.2	289	45
	20	35.5	70	66.3		10	338.0	321	57
	25	39.6	75	68.7		11	371.8	354	6
13	30	43.5	80	70.9	33.8	12	405.6	386	18
	35	46.9	85	73.1		13	439.4	418	30
	40	50.2	90	75.2		14	473.2	450	42
	45	53.2	95	77.3		15	507.0	482	54
	50	56.1	100	79.3		16	540.8	515	3
	5	19.1	55	63.3		7	274.4	261	21
	10	27.0	60	66.1		8	313.6	298	42
	15	33.1	65	68.9		9	352.8	336	—
	20	38.2	70	71.5		10	392.0	373	21
	25	42.7	75	74.0		11	431.2	410	42
14	30	46.8	80	76.4	39.2	12	470.4	448	—
	35	50.5	85	78.7		13	509.6	485	21
	40	54.0	90	81.0		14	548.8	522	42
	45	57.3	95	83.2		15	588.0	560	—
	50	60.4	100	85.4		16	627.2	597	21
	5	20.5	55	67.9		7	315.	300	—
	10	28.9	60	70.9		8	360.	342	54
	15	35.4	65	73.8		9	405.	385	45
	20	41.0	70	76.6		10	450.	428	36
	25	45.7	75	79.2		11	495.	471	27
15	30	50.1	80	81.8	45.	12	540.	514	18
	35	54.1	85	84.4		13	585.	557	9
	40	57.9	90	86.8		14	630.	600	—
	45	61.4	95	89.2		15	675.	642	54
	50	64.7	100	91.5		16	720.	685	45

SEE PAGE 42.

Pumps, Diameter in Inches.	Depth Shaft, Faths.	Diameter of the Cylinder.	Depth of Shaft in Fathoms.	Diameter of the Cylinder.	Gallons drawn at a 6 foot Stroke.	Strokes in 1 Min.	Gallons drawn in One Minute.	WATER drawn in one Hour. Hogfheads	Gal.
	5	21.9	55	72.4		7	358.4	341	21
	10	30.8	60	75.6		8	409.6	390	6
	15	37.8	65	78.7		9	460.8	438	54
	20	43.6	70	81.7		10	512.0	487	39
	25	48.8	75	84.5		11	563.2	536	24
16	30	53.4	80	87.3	51.2	12	614.4	585	9
	35	57.7	85	90.0		13	665.6	633	57
	40	61.7	90	92.6		14	716.8	682	42
	45	65.5	95	95.1		15	768.0	731	27
	50	69.0	100	97.6		16	819.2	780	12
	5	23.2	55	76.9		7	404.6	385	21
	10	32.8	60	80.3		8	462.4	440	24
	15	40.2	65	83.6		9	520.2	495	27
	20	46.4	70	86.8		10	578.0	550	30
	25	51.8	75	89.8		11	635.8	605	33
17	30	56.8	80	92.8	57.8	12	693.6	660	36
	35	61.3	85	95.6		13	751.4	715	39
	40	65.6	90	98.4		14	809.2	770	42
	45	69.6	95	101.1		15	867.0	825	45
	50	73.3	100	103.7		16	924.8	880	48
	5	24.6	55	81.4		7	453.6	432	—
	10	34.7	60	85.0		8	518.4	493	45
	15	42.5	65	88.5		9	583.2	555	27
	20	49.1	70	91.9		10	648.0	617	9
	25	54.9	75	95.1		11	712.8	678	54
18	30	60.1	80	98.3	64.8	12	777.6	740	36
	35	65.0	85	101.2		13	842.4	802	18
	40	69.4	90	104.1		14	907.2	864	—
	45	73.7	95	107.0		15	972.0	925	45
	50	77.6	100	109.8		16	1036.8	987	27

SEE PAGE 42.

Pumps, Diameter in Inches.	Depth Shaft, Faths.	Diameter of the Cylinder.	Depth of Shaft in Fathoms.	Diameter of the Cylinder.	Gallons drawn at a 6 foot Stroke.	Strokes in 1 Min.	Gallons drawn in One Minute.	WATER drawn in one Hour. Hogsheads	Gal
	5	25.9	55	86.0		7	505.4	481	21
	10	36.7	60	89.8		8	577.6	550	6
	15	44.9	65	93.4		9	649.8	618	54
	20	51.8	70	97.0		10	722.0	687	39
	25	57.9	75	100.4		11	794.2	756	24
19	30	63.5	80	103.7	72.2	12	866.4	825	9
	35	68.6	85	106.9		13	938.6	893	57
	40	73.3	90	110.0		14	1010.8	962	42
	45	77.7	95	113.0		15	1083.0	1031	27
	50	82.0	100	115.9		16	1155.2	1100	12
	5	27.2	55	90.5		7	561.4	534	42
	10	38.6	60	94.5		8	641.6	611	3
	15	47.2	65	98.4		9	721.8	687	27
	20	54.6	70	102.1		10	802.0	763	51
	25	61.0	75	105.7		11	882.2	840	12
20	30	66.8	80	109.1	80.2	12	962.4	916	36
	35	72.2	85	112.5		13	1042.6	992	60
	40	77.2	90	115.7		14	1122.8	1069	21
	45	81.8	95	118.9		15	1203.0	1145	45
	50	86.3	100	122.0		16	1283.2	1222	7
	5	28.6	55	95.0		7	618.8	589	21
	10	40.5	60	99.2		8	707.2	673	33
	15	49.6	65	103.3		9	795.6	757	45
	20	57.3	70	107.2		10	884.0	841	57
	25	64.1	75	110.9		11	972.4	926	6
21	30	70.2	80	114.6	88.4	12	1060.8	1010	18
	35	75.8	85	118.1		13	1149.2	1094	54
	40	81.0	90	121.5		14	1237.6	1178	42
	45	85.9	95	124.9		15	1326.0	1263	41
	50	90.6	100	128.1		16	1414.4	1347	3

SEE PAGE 42.

Pumps, Diameter in Inches.	Depth Shaft, Faths.	Diameter of the Cylinder.	Depth of Shaft in Fathoms.	Diameter of the Cylinder.	Gallons drawn at a 6 foot Stroke.	Strokes in 1 Min.	Gallons drawn in One Minute.	WATER drawn in one Hour. Hogsheads	Gal.
	5	30.0	55	99.5		7	677.6	645	21
	10	42.4	60	103.9		8	774.4	737	33
	15	52.0	65	108.2		9	871.2	829	45
	20	60.0	70	112.3		10	968.0	921	57
	25	67.1	75	116.2		11	1064.8	1014	6
22	30	73.5	80	120.0	96.8	12	1161.6	1106	18
	35	79.4	85	123.7		13	1258.4	1198	30
	40	84.9	90	127.2		14	1355.2	1290	42
	45	90.0	95	130.8		15	1452.0	1384	54
	50	94.9	100	134.2		16	1548.8	1475	3
	5	31.4	55	104.0		7	740.6	705	21
	10	44.4	60	108.7		8	846.4	806	6
	15	54.3	65	113.1		9	952.2	906	54
	20	62.7	70	117.3		10	1058.0	1007	39
	25	70.2	75	121.5		11	1163.8	1108	24
23	30	76.8	80	125.5	105.8	12	1269.6	1209	9
	35	83.0	85	129.4		13	1375.4	1309	57
	40	88.7	90	133.1		14	1481.2	1410	42
	45	94.1	95	136.8		15	1587.0	1511	27
	50	99.2	100	140.3		16	1692.8	1612	12
	5	32.8	55	108.6		7	806.4	768	—
	10	46.3	60	113.4		8	921.6	877	45
	15	56.7	65	118.0		9	1036.8	987	27
	20	65.5	70	122.5		10	1152.0	1097	9
	25	73.2	75	126.8		11	1267.2	1206	54
24	30	80.2	80	130.9	115.2	12	1382.4	1316	36
	35	86.6	85	135.0		13	1497.6	1426	18
	40	92.6	90	138.9		14	1612.8	1536	—
	45	98.2	95	142.7		15	1728.0	1645	45
	50	103.5	100	146.7		16	1843.2	1755	27

WEIGHT *and* QUANTITY *of* WATER *in* PUMPS.

SEE PAGE 42.

DIAMETER OF THE PUMPS.	Weight in lbs. of one Fathom deep of a Column of Water in each Pump, allowing 10.2 lbs. averdupoife to a Gallon.	DIAMETER OF THE PUMPS.	Weight in lbs. of one Fathom deep of a Column of Water in each Pump, allowing 10.2 lbs. averdupoife to a Gallon.	DIAMETER OF THE PUMPS.	Shews the Quantity of Water in Gallons, Ale Meafure, contained in one Foot deep of the Column in each Set of Pumps.	DIAMETER OF THE PUMPS.	Shews the Quantity of Water in Gallons, Ale Meafure, contained in one Foot deep of the Column in each Set of Pumps.
In.	*lbs.*	*In.*	*lbs.*	*In.*		*In.*	
5½	61.881	15	460.277	5½	1.0111	15	7.5208
6	73.644	15½	491.475	6	1.2033	15½	8.0306
6½	86.430	16	523.694	6½	1.4122	16	8.5570
7	100.238	16½	556.936	7	1.6378	16½	9.1002
7½	115.069	17	591.201	7½	1.8802	17	9.6601
8	130.922	17½	626.490	8	2.1392	17½	10.2367
8½	147.800	18	662.800	8½	2.4150	18	10.8300
9	165.699	18½	700.134	9	2.7078	18½	11.4401
9½	184.622	19	738.491	9½	3.0167	19	12.0668
10	204.567	19½	777.870	10	3.3426	19½	12.7085
10½	225.536	20	818.272	10½	3.6852	20	13.3704
11	247.527	20½	859.697	11	4.0445	20½	14.0453
11½	270.541	21	902.145	11½	4.4206	21	14.7409
12	294.578	21½	945.616	12	4.8133	21½	15.4512
12½	319.637	22	990.110	12½	5.2228	22	16.1782
13	345.720	22½	1035.626	13	5.6481	22½	16.9220
13½	372.825	23	1082.165	13½	6.0919	23	17.6824
14	400.953	23½	1129.727	14	6.5515	23½	18.4596
14½	430.104	24	1178.312	14½	7.0278	24	19.2534

SEE PAGE 42.

DIAMETER *of* CYLINDERS	Allowing seven pounds pressure upon every sq. Inch of the Cylinder, will counterpoise in pounds averdupoise.	DIAMETER *of* CYLINDERS	Allowing seven pounds pressure upon every sq. Inch of the Cylinder, will counterpoise in pounds averdupoise.	DIAMETER *of* CYLINDERS	Allowing 7 pounds pressure upon every sq. in. of the Cylinder, will counterpoise in pounds averdupoise.
8	351.85	33	5987.10	58	18494.59
9	445.32	34	6355.45	59	19137.84
10	549.78	35	6734.80	60	19792.08
11	665.23	36	7125.15	61	20457.31
12	791.68	37	7526.48	62	21133.54
13	929.12	38	7938.82	63	21820.76
14	1077.56	39	8362.15	64	22518.98
15	1237.00	40	8796.48	65	23228.20
16	1407.43	41	9241.80	66	23948.41
17	1588.86	42	9698.12	67	24679 62
18	1781.28	43	10165.43	68	25421.82
19	1984.70	44	10643 73	69	26175.02
20	2199.12	45	11133.04	70	26939.22
21	2424.53	46	11633.34	71	27714.40
22	2660.93	47	12144.64	72	28500.59
23	2908.33	48	12666.93	73	29297.77
24	3166.73	49	13200.21	74	30105.95
25	3436.12	50	13744.49	75	30925.12
26	3716.51	51	14299.77	76	31755.29
27	4007.89	52	14866.05	77	32596.45
28	4310.27	53	15443.32	78	33448.61
29	4623.65	54	16031.58	79	34311.76
30	4948.02	55	16630.84	80	35185.92
31	5283.38	56	17241.10	81	36071.05
32	5629.74	57	17862.35	82	36967.20

WEIGHT *of* CYLINDERS *Ten Feet Long, and in* PARTS.

SEE PAGE 47.

Weight of Lugs 8 in. sq. 1¼ deep 88lbs.	Flange 3½ in. broad, and 1⅛ thick.	Weight of cup ring 2 in. bd. ⅞ thick.	Weight of 1 ft. of Cylinder ⅞ thick.	WEIGHT of *CYLINDER.*		Weight of Lugs 8 in. sq. 2 deep, 132lbs.	Weight of Flan. 4 inches broad, 1½ thick	Cup Ring, 2 inches broad, 1 thick.	Weight of 1 ft. of Cylinder 1 thick.	WEIGHT of *CYLINDER.*	
Diam.	*lbs.*	*lbs.*	*lbs.*	*c.*	*q.*	*Diam.*	*lbs.*	*lbs.*	*lbs.*	*c.*	*q.*
16	68	28	143	14	2	46	253	81	457	45	0
17	71	29	152	15	1	47	258	83	467	45	3
18	74	31	161	16	0	48	263	84	476	46	3
19	77	32	169	16	3	49	267	86	486	47	3
20	80	34	178	17	2	50	272	87	496	48	2
21	84	36	186	18	2	51	277	89	506	49	2
22	87	37	194	19	1	52	282	91	515	50	2
23	90	38	203	20	0	53	287	92	525	51	2
24	93	39	212	20	3	54	292	94	535	52	1
25	96	41	220	21	2	55	297	96	544	53	1
Weight of Lugs 8 in. sq. 1½ deep 99lb.	Flange 3½ in. broad, and 1⅛ thick.	Weight of Cup Ring 2 in. bd. ⅞ thick.	Weight of 1 ft. of Cylinder ⅞ thick.	WEIGHT of *CYLINDER.*		Weight of Lugs 8 in. sq. 2½ deep 165lbs.	Flange 4 inches broad, and 1¾ thick.	Weight of Cup Ring 2 in. bd. 1⅛ thick	Weight of 1 ft. of Cylinder 1⅛ thick	WEIGHT of *CYLINDER.*	
26	100	42	228	22	2	56	353	110	625	61	1
27	103	43	237	23	1	57	359	112	636	62	2
28	106	45	246	24	0	58	364	113	647	63	2
29	109	46	254	24	3	59	370	115	658	64	2
30	112	48	263	25	3	60	376	117	669	65	2
31	116	49	271	26	2	61	382	119	680	66	2
32	119	51	280	27	1	62	387	121	691	67	3
33	122	52	288	28	1	63	393	122	702	68	3
34	125	53	297	28	3	64	398	124	712	69	3
35	128	55	305	29	3	65	404	126	723	70	3
Weight of Lugs 8 in. sq. 1¾ deep 116lbs.	Flange 3½ in. broad, and 1¼ thick.	Weight of Cup Ring 2 in. bd. ⅞ thick.	Weight of 1 ft. of Cylinder ⅞ thick.	WEIGHT of *CYLINDER.*		Weight of Lugs 8 in. sq. 3 deep, 198lbs.	Flange 4 inches broad, and 1¾ thick.	Weight of Cup Ring 2 in bd. 1⅛ thick	Weight of 1 ft. of Cylinder 1⅛ thick	WEIGHT of *CYLINDER.*	
36	146	56	314	30	3	66	410	128	734	72	0
37	149	58	322	31	2	67	416	130	745	73	1
38	153	59	331	32	2	68	421	132	756	74	1
39	157	60	339	33	1	69	427	133	767	75	1
40	160	62	348	34	0	70	433	135	778	76	1
41	164	63	356	34	3	71	438	137	789	77	1
42	168	65	365	35	3	72	444	139	800	78	1
43	171	66	373	36	2	73	449	141	811	79	1
44	175	68	382	37	1	74	455	143	822	80	2
45	179	69	390	38	0	75	461	144	833	81	2

WEIGHT *of One Foot length of* MALLEABLE IRON.

SQUARE IRON.		ROUND IRON.				CHAINS.			
Scantling.	WEIGHT.	DIAMETER.		CIRCUMFER.		Circumference of each Link.	Length of each Link.	WEIGHT.	
Inches.	Pounds.	Inches.	Weight in Pounds.	Inches.	Weight in Pounds.			Pounds.	Ounces.
1/4	0.21	1/4	0.16	1	0.26	7/16	1	0	3
3/8	0.47	3/8	0.37	1 1/4	0.41	1/2	1 1/4	0	3 1/2
1/2	0.84	1/2	0.66	1 1/2	0.59	9/16	1 1/2	0	4 1/4
5/8	1.34	5/8	1.03	1 3/4	0.82	5/8	1 3/4	0	5
3/4	1.89	3/4	1.48	2	1.05	11/16	2	0	5 3/4
7/8	2.57	7/8	2.02	2 1/4	1.34	3/4	2	0	6 1/2
1	3.36	1	2.63	2 1/2	1.65	13/16	2 1/4	0	7 1/4
1 1/8	4.25	1 1/8	3.33	2 3/4	2.01	7/8	2 1/4	0	8
1 1/4	5.25	1 1/4	4.12	3	2.37	15/16	2 1/2	0	8 3/4
1 3/8	6.35	1 3/8	4.98	3 1/4	2.79	1	2 3/4	0	9 1/2
1 1/2	7.56	1 1/2	5.93	3 1/2	3.24	1 1/16	3	0	10 1/4
1 5/8	8.87	1 5/8	6.96	3 3/4	3.69	1 1/8	3 1/4	0	11 1/4
1 3/4	10.29	1 3/4	8.08	4	4 23	1 3/16	3 1/4	0	12 1/2
1 7/8	11.81	1 7/8	9.27	4 1/2	5 35	1 1/4	3 1/2	0	14
2	13.44	2	10.55	5	6.61	1 5/16	3 1/2	1	—
2 1/4	17.01	2 1/4	13.35	5 1/2	7.99	1 3/8	3 1/2	1	2
2 1/2	21.	2 1/2	16.48	6	9.51	1 7/16	3 1/2	1	4
2 3/4	25.41	2 3/4	19.95	6 1/2	11.18	1 1/2	3 1/2	1	6
3	30.24	3	23.73	7	12.96	1 9/16	3 3/4	1	8
3 1/2	41.16	3 1/4	27.85	7 1/2	14.78	1 5/8	3 3/4	1	10
4	53.76	3 1/2	32.32	8	16.92	1 11/16	3 3/4	1	12 1/2
4 1/2	68.04	3 3/4	37.09	8 1/2	19.21	1 3/4	3 3/4	1	15 1/2
5	84.00	4	42.21	9	21.53	1 13/16	3 3/4	2	3 1/2
6	120.96	4 1/2	53.41	10	26.43	1 7/8	3 3/4	2	8 1/2
7	164.64	5	65.93	11	31.99	1 15/16	4	2	14 1/2

WEIGHT *of* FLAT *and* ROLLED IRON, *One Foot long.*

Broad.	Thick.	Weight in Pounds.
1/2	1/8	0.21
	3/16	0.31
	1/4	0.42
	3/8	0.63
3/4	1/8	0.31
	3/16	0.47
	1/4	0.63
	3/8	0.94
	1/2	1.26
	5/8	1 57
1	1/8	0.42
	3/16	0.63
	1/4	0.84
	3/8	1.26
	1/2	1.68
	5/8	2.10
	3/4	2.52
	7/8	2.94
1 1/4	1/8	0.52
	3/16	0.78
	1/4	1.05
	3/8	1.57
	1/2	2 10
	5/8	2.62
	3/4	3.15
	7/8	3.67
	1	4.20
	1 1/8	4.72
1 3/8	1/8	0.57
	3/16	0.86
	1/4	1.18

Broad.	Thick.	Weight in Pounds.
1 3/8	3/8	1.73
	1/2	2.31
	5/8	2.88
	3/4	3.46
	7/8	4.04
	1	4.62
	1 1/8	5.19
	1 1/4	5.77
1 1/2	1/8	0.63
	3/16	0.94
	1/4	1.26
	3/8	1.89
	1/2	2.52
	5/8	3.15
	3/4	3.78
	7/8	4.41
	1	5.04
	1 1/8	5.67
	1 1/4	6.30
1 3/4	1/8	0.73
	3/16	1.10
	1/4	1.47
	3/8	2.20
	1/2	2.94
	5/8	3.67
	3/4	4.41
	7/8	5.14
	1	5.87
	1 1/8	6.60
	1 1/4	7.35
	1 3/8	8.07
	1 1/2	8.80

Broad.	Thick.	Weight in Pounds.
2	1/8	0.84
	3/16	1.26
	1/4	1.68
	3/8	2.52
	1/2	3.36
	5/8	4.20
	3/4	5 04
	7/8	5.88
	1	6.72
	1 1/8	7.56
	1 1/4	8.40
	1 3/8	9.24
	1 1/2	10.08
2 1/4	1/8	0.94
	3/16	1.41
	1/4	1.89
	3/8	2.83
	1/2	3.78
	5/8	4.72
	3/4	5.66
	7/8	6.61
	1	7.56
	1 1/8	8.50
	1 1/4	9.45
	1 3/8	10.39
	1 1/2	11.34
	1 3/4	13.22
	2	15.12
2 1/2	1/8	1.05
	3/16	1.57
	1/4	2.10

Broad.	Thick.	Weight in Pounds.
2½	3/8	3.15
	1/2	4.20
	5/8	5.25
	3/4	6.30
	7/8	7.35
	1	8.40
	1⅛	9.55
	1¼	10.50
	1½	12.60
	2	16.80
2¾	1/8	1.15
	3/16	1.73
	1/4	2.31
	3/8	3.46
	1/2	4.62
	5/8	5.77
	3/4	6.93
	7/8	8.08
	1	9.24
	1⅛	10.39
	1¼	11.55
	1½	13.86
	2	18.48
	2½	23.10
3	1/8	1.26
	3/16	1.89
	1/4	2.52
	3/8	3.78
	1/2	5.04
	5/8	6.30
	3/4	7.56
	7/8	8.82
	1	10.08
	1⅛	11.34
	1¼	12.60
	1½	15.12
	2	20.16
	2½	25.20
3¼	1/8	1.36
	3/16	2.04
	1/4	2.73
	3/8	4.09
	1/2	5.46
	5/8	6.82
	3/4	8.19
	7/8	9.55
	1	10.92
	1⅛	12.28
	1¼	13.65
	1½	16.38
	2	21.84
	2½	27.39
	3	32.76
3½	1/8	1.47
	3/16	2.20
	1/4	2.94
	3/8	4.41
	1/2	5.88
	5/8	7.35
	3/4	8.82
	7/8	10.29
	1	11.76
	1¼	14.70
	1½	17.64
	2	23.52
	2½	29.40
	3	35.28
3¾	1/8	1.57
	3/16	2.36
	1/4	3.15
	3/8	4.72
	1/2	6.30
	5/8	7.87
	3/4	9.45
	7/8	11.02
	1	12.60
	1¼	15.75
	1½	18.90
	2	25.20
	2½	31.50
	3	37.80
4	1/8	1.68
	3/16	2.52
	1/4	3.36
	3/8	5.04
	1/2	6.72
	5/8	8.40
	3/4	10.08
	7/8	11.76
	1	13.44
	1¼	16.80
	1½	20.18
	2	26.88
	2½	33.65
	3	40.32
	3½	47.04

DIMENSIONS *of* BOILER PLATES.

SEE PAGE 43.

Diameter of Boilers.	BOTT. PLATES.					LAGGON PLATES					FLUE PLATES.					TOP PLATES.					TOP PLATES.					*Bottom Crown.*		*Top Crown.*	
	Num. of Plates.	Length of Plates.		Br. at one end.	Br. at the other.	Num. of Plates.	Length of Pla.es.		Br. at one end.	Br. at the other.	Num. of Plates.	Length of Plates.		Br. at one end.	Br. at the other.	Num. of Plates.	Length of Plates		Br. at one end.	Br. at the other	Num. of Plates.	Length of Plates.		Br. at one end.	Br. at the other.	Diameter.		Diameter.	
Feet.		Feet.	Inches	Inches.	Inches.		Feet.	Inches.	Inches.	Inches.		Feet.	Inches.	Inches.	Inches.		Feet.	Inches.	Inches.	Inches.		Feet.	Inches.	Inches.	Inches.	Feet.	Inches.	Feet.	Inches.
4	11		9⅝	12	9¼	—	0	—	—	—	11	1	3¾	15¼	12	11	2	0¾	15¼	9¼	—	0	—	—	—	2	5½	2	5½
4½	12	1	0¼	12½	8⅝	—	0	—	—	—	12	1	5½	15⅝	12½	12	2	5¾	15⅝	8⅝	—	0	—	—	—	2	5½	2	5½
5	13	1	4	14¼	8¼	—	0	—	—	—	13	1	7¼	16	14¼	13	2	10	16	8¼	—	0	—	—	—	2	5½	2	5½
5½	14	1	7¼	14⅝	7¾	—	0	—	—	—	14	1	9	16¼	14⅝	14	3	3½	16¼	7¾	—	0	—	—	—	2	5½	2	5½
6	15	1	9	14⅞	7¼	—	0	—	—	—	15	1	10¾	16½	14⅞	15	3	7½	16½	7¼	—	0	—	—	—	2	5½	2	5½
6½	17	2	0½	14¼	6⅝	—	0	—	—	—	17	2	0½	15⅞	14¼	17	4	0½	15⅞	6⅝	—	0	—	—	—	2	5½	2	5½
7	18	2	3½	14½	6⅜	—	0	—	—	—	18	2	2½	16⅛	14½	18	2	3½	16⅛	13½	18	2	3½	13½	6⅜	2	5½	2	5½
7½	19	2	6	14¾	6⅛	—	0	—	—	—	19	2	4½	16⅜	14¾	19	2	6	16⅜	13½	19	2	6	13½	6⅛	2	5½	2	5½
8	20	2	8¾	14⅞	6	—	0	—	—	—	20	2	5	16½	14⅞	20	2	8½	16⅝	13⅝	20	2	8½	13⅝	6	2	5½	2	5½
8½	21	3	0¾	15⅛	5¾	—	0	—	—	—	21	2	6¾	16¾	15⅛	21	2	10¾	16¾	13⅝	21	2	10¾	13⅝	5¾	2	5½	2	5½
9	23	3	3½	14¾	5⅜	—	0	—	—	—	23	2	8½	16¼	14¾	23	3	1	16¼	13¼	23	3	1	13¼	5⅜	2	5½	2	5½
9½	24	3	0½	14⅞	6¼	—	0	—	—	—	24	2	11½	16⅜	14⅞	24	3	1½	16⅜	13⅜	24	3	1½	13⅜	6¼	3	1½	3	1½
10	25	3	3¼	15	6	—	0	—	—	—	25	3	1¼	16½	15	25	3	3½	16½	13⅞	25	3	3½	13⅞	6	3	1½	3	1½
10½	26	3	6¼	15⅛	5⅞	—	0	—	—	—	26	3	3	16¾	15⅛	26	3	5½	16¾	14	26	3	5½	14	5⅞	3	1½	3	1½
11	22	2	5½	14¼	6⅝	28	1	6½	14¾	11½	31	3	3	14⅞	13½	27	3	5	16⅞	14¼	23	3	5	16⅜	7⅞	3	1½	4	1½
11½	23	2	7½	14⅛	6¾	30	1	7½	14⅜	11¼	33	3	4½	14¾	13¼	28	3	7½	17	14¼	24	3	7½	16⅜	7⅝	3	1½	4	1½
12	24	2	9	14	6¼	31	1	8½	14½	11¼	34	3	6	14⅞	13⅜	30	3	10	16⅝	13⅞	25	3	10	16¼	7⅜	3	1½	4	1½
12½	24	2	8	14½	7	32	1	9½	14¾	11⅜	36	3	7¾	14⅝	13¼	31	4	0	16¾	13⅞	26	4	0	16¼	7⅛	3	7½	4	1½
13	25	2	9½	14⅜	6¾	34	1	10½	14½	11	43	3	9½	12⅞	11⅞	32	4	2½	16⅞	13⅞	27	4	2½	16⅛	7	3	7½	4	1½
13½	26	2	11½	14¼	6⅝	35	1	11½	14⅝	11	44	3	11	13	11⅞	33	4	4½	17	13⅞	28	4	4½	16⅛	6⅞	3	7½	4	1½
14	26	3	1½	14¾	6⅝	36	2	1	14⅝	11⅛	45	4	0½	13¼	12	35	4	6½	16⅝	13⅝	29	4	6½	16⅛	6⅝	3	7½	4	1½
14½	27	3	0½	14⅝	7	38	2	2½	14½	11	46	4	2	13½	12¼	36	4	9	16¾	13⅝	30	4	9	16	6½	4	1½	4	1½
15	28	3	2	14½	6⅞	39	2	3½	14½	11	48	4	3½	13¼	12	37	4	11½	16⅞	13¾	31	4	11½	16	6⅜	4	1½	4	1½
15½	29	3	4	14½	6¾	40	2	5	14⅝	11	49	4	4½	13⅜	12⅛	38	5	2½	17	13¾	32	5	2½	16	6¼	4	1½	4	1½
16	30	3	6	14½	6½	41	2	7	14¾	11⅛	51	4	6½	13¼	12⅛	40	5	5½	16⅝	13½	33	5	5½	16	6⅛	4	1½	4	1½
16½	31	3	8	14½	6⅜	42	2	8½	14¾	11⅛	52	4	8½	13½	12¼	41	5	7½	16⅝	13⅜	34	5	7½	15⅞	6	4	1½	4	1½
17	32	3	10	14⅝	6¼	43	2	10	14⅞	11¼	53	4	10½	13⅝	12⅜	42	5	9½	16¾	13½	34	5	9½	16⅜	6	4	1½	4	1½

SEE PAGE 44.

Diameter of Boilers.	Diameter of the bottom Crown Plates.		Diameter at the bottom of the Flue Plates. Page 38.		Diameter of the top of the Crown Plate, meaſuring from pt. to pt.		Length of bottom Plates, meaſuring round Laggon.		Length of Flue Plates.		Length of top Plates.		Height the Boiler ſtands.		Riſe of Crown.	Radius of the bottom.		Wt. of Plates: Thickneſs to Flues, 1/4; Top Plates, both rows, 3/16			Wt. of Plates: Thickneſs to Flues, 3/8; Firſt Top Courſe, 1/4; Second Top Courſe, 3/16			Weight of Rabbit Iron.	
Feet.	Feet.	Inches	Feet.	Inches.	Feet.	Inches.	Feet.	Inches.	Feet.	Inches.	Feet.	Inches.	Feet.	Inches.	Inches.	Feet.	Inches.	Cwt.	Qrs.	Pounds.	Cwt.	Qrs.	Pounds	Cwt.	Qrs.
4	2	4	3	6	2	4		8⅛	1	2¼	1	11¼	3	3	4	4	1½	4	3	10	6	2	6	0	3
4½	2	4	3	11½	2	4		10¾	1	4	2	4¼	3	8	4¾	4	4¾	5	3	21	8	1	9	1	0
5	2	4	4	5	2	4	1	2½	1	5¾	2	7	4	1	5½	4	9	7	1	22	10	1	26	1	1
5½	2	4	4	10½	2	4	1	5¾	1	7½	2	11½	4	6	6¼	5	0¼	9	0	0	12	2	23	1	2
6	2	4	5	4	2	4	1	7½	1	9¼	3	4	4	11	6¾	5	3¾	10	1	17	14	2	23	1	3
6½	2	4	5	9¼	2	4	1	11	1	11	3	8½	5	4	7½	5	7	12	1	6	17	1	22	2	0
7	2	4	6	2½	2	4	2	2	2	1	2	1¼	5	9	8	5	10½	14	2	7	19	3	13	2	1
7½	2	4	6	8	2	4	2	4½	2	2	2	3½	6	2	8½	6	1¾	16	2	18	22	3	0	2	2
8	2	4	7	1	2	4	2	7¼	2	3½	2	6	6	7	9	6	5	18	2	16	25	1	19	2	3
8½	2	4	7	7	2	4	2	10¾	2	5¼	2	8	6	11	9¼	6	9	21	0	3	28	2	12	3	0
9	2	4	8	1	2	4	3	1	2	7	2	10½	7	4	9½	7	2	23	2	20	32	1	18	3	2
9½	3	0	8	6	3	0	2	11	2	10	2	11	7	8	10½	7	7	26	0	14	35	2	19	4	0
10	3	0	8	11½	3	0	3	1¾	2	11¾	3	1¼	8	1	11	8	0	28	3	6	39	1	13	4	2
10½	3	0	9	5	3	0	3	4¾	3	1½	3	3½	8	6	12	8	4	31	2	20	43	1	11	5	0
11	3	0	9	10	4	0	2	4	3	1½	3	3½	8	11	12¾	8	8½	35	0	10	47	2	13	5	2
11½	3	0	10	3	4	0	2	6	3	3	3	5½	9	4	13¼	9	1	38	0	4	51	3	18	6	0
12	3	0	10	8	4	0	2	7½	3	4½	3	8	9	9	13¾	9	6	41	1	13	56	1	23	6	2
12½	3	6	11	2	4	0	2	6½	3	6¼	3	10	10	1½	14¾	9	10	44	2	24	61	0	8	6	3
13	3	6	11	8	4	0	2	8	3	8	4	0½	10	6	15½	10	3	48	1	16	66	0	21	7	0
13½	3	6	12	1	4	0	2	10	3	9½	4	2½	10	10½	16¼	10	8	51	3	25	71	0	5	7	2
14	3	6	12	6	4	0	3	0	3	11	4	4½	11	3½	16¾	11	0	55	2	5	76	0	11	8	0
14½	4	0	13	0	4	0	2	11	4	0½	4	7	11	8	17½	11	5	59	3	4	82	0	0	8	2
15	4	0	13	6	4	0	3	0½	4	2	4	9½	12	1½	18¼	11	10	63	3	2	87	1	12	9	0
15½	4	0	13	10	4	0	3	2½	4	3	5	0	12	6	19	12	4	68	1	0	93	2	2	9	2
16	4	0	14	4	4	0	3	4½	4	5	5	2	13	0	19½	12	8	73	1	23	100	0	22	9	3
16½	4	0	14	9½	4	0	3	6½	4	7	5	5	13	6	20	13	0	77	3	21	107	0	6	10	0
17	4	0	15	3	4	0	3	8½	4	9	5	8	14	0	20½	13	4	82	3	9	113	3	16	10	2

WEIGHT *of* CAST IRON PIPES *or* TREES.

Diameter of Pumps.	*Metal $\frac{5}{8}$ thick.*					*Metal $\frac{3}{4}$ thick.*					*Metal $\frac{7}{8}$ thick.*					*Metal* 1 *in. thick.*				
	Wt. of flanges $\frac{3}{4}$ thick no beads.	Wt. of 1 foot long of Pump	Weight of Pump 9 feet long.			Wt. of flange $2\frac{1}{2}$ bd. $\frac{3}{4}$ th. no beads.	Wt. of 1 foot long of Pump	Weight of Pump 9 feet long.			Wt. of flanges 3 bd. 1 th. no beads.	Wt. of 1 foot long of Pump	Weight of Pump 9 feet long			Wt. of flanges 3 bd. 1 in. th. no beads	Wt. of 1 foot long of Pump	Weight of Pump 9 feet long.		
In.	*lbs.*	*lbs.*	*c.*	*q.*	*lbs.*	*lbs.*	*lbs.*	*c.*	*q.*	*lbs.*	*lbs.*	*lbs.*	*c.*	*q.*	*lbs*	*lbs.*	*lbs.*	*c.*	*q.*	*lbs.*
4	23.3	28.1	2	1	24	24.1	34.6	2	3	27	00.0	00.0	0	0	00	00.0	00.0	0	0	00
4½	25.0	31.0	2	2	25	25.7	38.1	3	1	4	00.0	00.0	0	0	00	00.0	00.0	0	0	00
5	26.5	34.2	2	3	26	27.3	41.9	3	2	12	00.0	00.0	0	0	00	00.0	00.0	0	0	00
5½	28.1	37.3	3	1	0	29.0	45.6	3	3	19	00.0	00.0	0	0	00	00.0	00.0	0	0	00
	Flanges one Inch thick.					Flanges one Inch thick.					Flanges one Inch thick.					Flanges one Inch thick.				
6	39.5	40.4	3	2	11	40.5	49.2	4	1	6	41.5	58.4	5	0	7	42.5	68.0	5	3	10
6½	41.5	43.4	3	3	12	42.5	52.9	4	2	14	43.5	62.9	5	1	21	44.5	72.9	6	1	0
7	43.5	46.4	4	0	13	44.5	56.5	4	3	21	45.5	67.1	5	3	5	46.5	77.8	6	2	18
7½	45.5	49.4	4	1	14	46.5	60.1	5	0	27	47.5	71.3	6	0	17	48.5	82.5	7	0	7
8	47.6	52.4	4	2	15	48.6	63.7	5	2	6	49.6	75.5	6	2	1	50.6	87.3	7	1	24
8½	49.6	55.4	4	3	16	50.7	67.3	5	3	12	51.7	79.7	6	3	13	52.7	92.1	7	3	12
	Metal $\frac{3}{4}$ thick, (two beads,) Flanges 3 in. broad.					Metal $\frac{7}{8}$ thick. (two beads,) Flanges 3 in. broad.					Metal 1 In. thick. (two beads,) Flanges 3 in. broad.					Metal $1\frac{1}{8}$ thick. (two beads,) Flan. 3 in. bd. $1\frac{1}{4}$ thick.				
9	83.0	71.0	6	1	22	86.0	84.0	7	2	2	88.0	97.1	8	2	10	107.8	111.2	9	3	16
9½	87.0	74.5	6	3	1	88.5	88.2	7	3	14	90.0	102.0	9	0	0	110.9	116.5	10	1	11
10	89.5	78.7	7	0	10	91.0	92.5	8	0	27	92.2	106.9	9	1	18	114.0	121.8	10	3	6
10½	92.2	82.7	7	1	18	93.5	96.8	8	2	12	94.7	111.8	9	3	8	117.2	127.2	11	1	2
11	95.1	86.4	7	2	25	95.8	101.0	8	3	25	97.5	116.7	10	0	26	120.3	132.6	11	2	25
11½	98.2	90.0	8	0	4	98.0	105.3	9	1	9	100.5	121.6	10	2	17	123.4	138.0	12	0	21
12	101.2	93.0	8	1	12	102.5	109.8	9	2	22	103.8	126.4	11	0	8	126.6	143.5	12	2	18
12½	103.8	96.6	8	2	20	105.8	114.0	10	0	8	107.0	131.2	11	2	0	129.7	149.1	13	0	15
13	106.7	100.2	9	0	0	109.2	118.1	10	1	22	110.2	136.1	11	3	19	132.8	154.6	13	2	12
13½	108.6	103.8	9	1	6	111.8	122.3	10	3	7	113.0	140.9	12	1	10	136.0	160.0	14	0	8
14	109.2	107.5	9	2	12	114.5	126.6	11	0	22	115.7	145.8	12	3	0	139.1	165.2	14	2	4
14½	110.7	111.1	9	3	18	117.3	130.9	11	2	7	118.5	150.6	13	0	18	142.3	170.9	15	0	0
15	111.8	114.8	10	0	25	120.1	135.1	11	3	20	121.2	155.5	13	2	9	145.5	176.4	15	1	25
15½	113.2	118.4	10	2	3	122.2	139.4	12	1	5	123.3	160.4	13	3	27	148.6	181.9	15	3	22
16	114.7	122.0	10	3	8	124.3	143.6	12	2	17	125.5	165.3	14	1	17	151.6	187.3	16	1	18
16½	116.0	126.5	11	0	14	126.4	147.9	13	0	2	128.1	170.2	14	3	7	154.7	192.8	16	3	14
17	117.1	129.3	11	1	21	128.5	152.1	13	1	13	130.7	175.0	15	0	26	157.8	198.3	17	1	10
17½	118.5	132.9	11	3	0	130.7	156.4	13	2	26	132.8	179.9	15	2	16	160.9	203.7	17	3	6
18	120.0	136.7	12	0	6	133.0	160.6	14	0	10	135.0	184.8	16	0	6	164.0	209.3	18	1	3
19	000.0	000.0	00	0	00	137.5	169.1	14	3	7	139.7	194.5	16	3	24	170.2	220.2	19	0	24
20	000.0	000.0	00	0	00	142.0	177.6	15	2	1	144.4	204.2	17	2	22	176.4	231.1	20	0	16
21	000.0	000.0	00	0	00	146.5	186.1	16	1	1	149.1	214.0	18	2	11	182.6	242.0	21	0	8
22	000.0	000.0	00	0	00	151.0	194.6	16	3	26	153.8	224.6	19	1	19	188.8	252.9	22	0	1
23	000.0	000.0	00	0	00	155.5	203.1	17	2	23	158.8	233.3	20	0	18	195.0	263.9	22	3	22
24	000.0	000.0	00	0	00	161.0	211.6	18	1	20	163.3	243.1	20	3	27	201.2	274.9	23	3	15

WEIGHT *of* WIND BORES, *(square Bottoms)* 8 *feet long.*

Plate 5, *Fig.* 35.—*Page* 58.

Metal one Inch and three-eights thick, with Holes in them; ſwelled part thirty Inches long, and ſwelled two Inches on each ſide.

DIAMETER. Inches.	WEIGHT. Cwt.	WEIGHT. Qrs.
6	8	0
6½	8	1
7	8	3
7½	9	1
8	9	3
8½	10	1
9	10	3
9½	11	1
10	11	3
10½	12	1
11	12	2
11½	13	0
12	13	2
12½	14	0
13	14	2
13½	15	0
14	15	2
14½	16	0
15	16	2
15½	17	0
16	17	2
16½	18	0
17	18	2
17½	19	0
18	19	3
18½	20	1
19	20	3

WEIGHT of Ten Yards in Length of ROPE.—(Circumference given in Inches.)

(Six Inches and three quarters Shroud laid Rope is equal to a ſeven Inches Cable laid Rope, and all above ſeven Inches are cable laid.)

Circum.	Wt. in pounds	Circum.	Wt. in pounds	Circum.	Wt in pounds	Circum.	Wt. in pounds	Circum.	Wt. in pounds	Circum.	Wt. in pounds	Circum.	Wt. in pounds	Circum.	Wt. in pounds
1	1.00	2	4.4	3¼	12.0	4½	22.8	5¾	37.0	7	53.0	8¼	74.1	9½	97.6
1¼	1.75	2¼	5.6	3½	14.0	4¾	25.3	6	41.0	7¼	57.4	8½	78.6	9¾	102.6
1½	2.60	2½	7.0	3¾	16.1	5	28.0	6¼	45.0	7½	61.4	8¾	83.2	10	107.7
1¾	3.25	2¾	8.5	4	18.2	5¼	30.8	6½	49.0	7¾	65.5	9	87.9	10¼	112.9
1⅞	3.40	3	10.2	4¼	20.4	5½	33.8	6¾	53.0	8	69.7	9¼	92.7	10½	118.2

WIDTH of the two Rows of Top Plates of BOILERS, in the intermediate parts.

IN INCHES.—SEE PAGE 43.

B. Feet.	FIRST COURSE OF TOP PLATES.							SECOND COURSE OF TOP PLATES.						
Diam.	Width at end.	Second width.	Third width.	Fourth width.	Fifth width.	Sixth width.	Seventh width.	Width at end.	Second width.	Third width.	Fourth width.	Fifth width.	Sixth width.	Seventh width.
6	16½	16.2	14.9	12.9	10.4	7¼	00.0	0	0.0	0.0	0.0	0.0	0.0	0
6½	15⅞	15.5	14.3	12.3	9.7	6⅝	00.0	0	0.0	0.0	0.0	0.0	0.0	0
7	16⅛	15.9	15.4	14.6	13½	0.0	00.0	13½	11.9	10.2	8.4	6⅜	0.0	0
7½	16⅜	16.2	15.6	14.7	13½	0.0	00.0	13½	12.0	10.2	8.2	6⅛	0.0	0
8	16⅝	16.5	15.8	14.9	13⅝	0.0	00.0	13⅝	12.0	10.2	8.1	6.0	0.0	0
8½	16¾	16.6	16.0	15.0	13⅝	0.0	00.0	13⅝	12.0	10.1	8.0	5¾	0.0	0
9	16¼	16.0	15.5	15.4	13¼	0.0	00.0	13¼	11.5	9.6	7.6	5⅜	0.0	0
9½	16⅜	16.3	15.7	14.8	13⅜	0.0	00.0	13⅜	12.0	10.3	8.3	6¼	0.0	0
10	16½	16.5	15.8	14.9	13⅞	0.0	00.0	13⅞	12.0	10.2	8.1	6.0	0.0	0
10½	16¾	16.5	16.0	15.1	14.0	0.0	00.0	14	12.0	10.1	8.0	5⅞	0.0	0
11	16⅞	16.7	16.3	15.7	14.9	14¼	00.0	16⅜	14.7	13.0	11.2	9.2	7⅞	0
11½	17	16.8	16.4	15.8	15.0	14¼	00.0	16⅜	14.6	12.9	11.1	9.1	7⅝	0
12	16⅝	16.4	16.1	15.6	14.6	13⅞	00.0	16¼	14.5	12.8	10.9	8.8	7⅜	0
12½	16¾	16.5	16.2	15.5	14.6	13⅞	00.0	16¼	14.5	12.8	10.8	8.7	7⅛	0
13	16⅞	16.6	16.2	15.8	14.7	13⅞	00.0	16⅛	14.5	12.7	10.7	8.9	7.0	0
13½	17	16.8	16.6	16.2	15.7	15.0	13⅝	16⅛	14.9	13.5	12.0	10.4	8.7	6⅞
14	16⅝	16.5	16.0	15.8	15.2	14.4	13⅝	16⅛	14.8	13.4	11.9	10.3	8.5	6⅝
14½	16¾	16.5	16.4	16.0	15.4	14.6	13⅝	16	14.7	13.4	11.8	10.2	8.4	6½
15	16⅞	16.7	16.4	16.0	15.4	14.6	13¾	16	14.7	13.4	11.7	10.1	8.3	6⅜
15½	17	16.8	16.5	16.0	15.5	14.6	13¾	16	14.6	13.2	11.6	10.0	8.2	6¼
16	16⅝	16.5	16.2	15.7	15.1	14.4	13½	16	14.6	13.2	11.5	9.8	8.1	6⅛
16½	16⅝	16.6	16.3	15.8	15.2	14.5	13⅜	15⅞	14.6	13.1	11.4	9.7	7.8	6
17	16¾	16.7	16.4	16.0	15.4	14.4	13½	16⅜	14.9	13.6	11.7	9.9	8.0	6

QUANTITY *and* WEIGHT *of* COAL *in a Statute* ACRE.

Height of the Bed.	NEWCASTLE MEASURE.			Waggons,	Tons,	Stacks,	Stacks,
				95 feet, or	20 Cwt.	56 solid	60 solid
F. In.	Boles.	Chaldrons.	Tens of 440 Boles, or 55 Cubic Yards	20 Boles.	each.	Feet.	Feet.
1. 6	19360	806.6	44.0	968.0	2286	1613	1505
7	20436	851.5	46.4	1021.8	2413	1703	1589
8	21511	896.3	48.8	1075.5	2540	1792	1672
9	22586	941.1	51.3	1129.3	2667	1882	1756
10	23662	985.9	53.7	1183.1	2794	1971	1839
11	24737	1030.7	56.2	1236.8	2921	2061	1923
2. 0	25813	1075.5	58.6	1290.6	3048	2151	2006
1	26888	1120.3	61.1	1344.4	3175	2240	2090
2	27964	1165.1	63.5	1398.2	3302	2330	2174
3	29040	1209.9	66.0	1452.0	3429	2419	2257
4	30115	1254.7	68.5	1505.7	3556	2509	2341
5	31191	1299.5	70.8	1559.5	3683	2599	2424
6	32266	1344.4	73.3	1613.3	3810	2688	2508
7	33342	1389.2	75.7	1667.1	3937	2778	2592
8	34417	1434.0	78.2	1720.8	4064	2868	2676
9	35493	1478.8	80.6	1774.6	4191	2957	2759
10	36568	1523.6	83.1	1828.4	4318	3047	2843
11	37644	1568.4	85.5	1882.2	4445	3136	2926
3. 0	38720	1613.3	88.0	1936.0	4572	3226	3010
1	39795	1658.1	90.4	1989.7	4699	3316	3094
2	40871	1702.9	92.8	2043.5	4826	3405	3178
3	41946	1747.7	95.3	2097.3	4953	3495	3261
4	43022	1792.5	97.7	2151.0	5080	3585	3345
5	44097	1837.3	100.2	2204.8	5207	3674	3429
6	45173	1882.2	102.6	2258.6	5334	3764	3512
7	46248	1927.0	105.1	2312.4	5461	3854	3596
8	47324	1971.8	107.5	2366.2	5588	3943	3680
9	48400	2016.6	110.0	2420.0	5715	4033	3764
10	49475	2061.4	112.4	2473.8	5842	4122	3847
11	50551	2106.2	114.8	2527.5	5969	4212	3930
4. 0	51626	2151.1	117.3	2581.3	6096	4302	4014
1	52702	2195.9	119.7	2635.1	6223	4391	4098
2	53777	2240.7	122.2	2688.8	6350	4481	4181
3	54853	2285.5	124.6	2742.6	6477	4571	4265
4	55928	2330.3	127.1	2796.4	6604	4660	4349
5	57004	2375.1	129.5	2850.2	6731	4750	4432
6	58080	2420.0	132.0	2904.0	6858	4840	4516
7	59155	2464.8	134.4	2957.7	6985	4929	4600
8	60231	2509.6	136.8	3011.5	7112	5019	4684
9	61306	2554.4	139.3	3065.3	7239	5108	4767

Height of the Bed.	NEWCASTLE MEASURE.			Waggons, 95 feet, or 20 Boles.	Tons, 20 Cwt. each.	Stacks, 56 solid Feet.	Stacks, 60 solid Feet.
F. In.	Boles.	Chaldrons.	Tens of 440 Boles, or 55 Cubic Yards				
4.10	62382	2599.2	141.7	3119.1	7366	5198	4851
11	63457	2644.0	144.2	3172.8	7493	5288	4935
5. 0	64533	2688.8	146.6	3226.6	7620	5377	5018
1	65608	2733.6	149.1	3280.4	7747	5467	5102
2	66684	2778.4	151.5	3334.2	7874	5556	5185
3	67760	2823.2	154.0	3388.0	8001	5646	5269
4	68835	2868.0	156.4	3441.7	8128	5736	5352
5	69911	2912.8	158.8	3495.5	8255	5825	5436
6	70986	2957.7	161.3	3549.3	8382	5915	5519
7	72062	3002.5	163.7	3603.1	8509	6005	5603
8	73137	3047.3	166.2	3656.8	8636	6094	5686
9	74213	3092.1	168.6	3710.6	8763	6184	5770
10	75288	3136.9	171.1	3764.4	8890	6273	5854
11	76364	3181.7	173.5	3818.2	9017	6363	5937
6. 0	77440	3226.6	176.0	3872.0	9144	6453	6021
1	78515	3271.4	178.4	3925.7	9271	6542	6105
2	79591	3316.2	180.8	3979.5	9398	6632	6188
3	80666	3361.0	183.3	4033.3	9525	6722	6272
4	81742	3405.8	185.7	4087.0	9652	6811	6356
5	82817	3450.6	188.2	4140.8	9779	6901	6439
6	83893	3495.4	190.6	4194.6	9906	6990	6522
7	84968	3540.2	193.1	4248.3	10033	7080	6606
8	86044	3585.0	195.5	4302.0	10160	7170	6690
9	87119	3629.8	198.0	4355.8	10287	7259	6774
10	88195	3674.6	200.4	4409.6	10414	7349	6858
11	89270	3719.5	202.8	4463.4	10541	7439	6941
7. 0	90346	3764.4	205.3	4517.2	10668	7528	7024
1	91421	3809.2	207.7	4571.0	10795	7618	7108
2	92497	3854.0	210.2	4624.8	10922	7708	7192
3	93572	3899.8	212.6	4678.6	11049	7797	7276
4	94648	3943.6	215.1	4732.4	11176	7887	7360
5	95724	3988.4	217.5	4786.2	11303	7976	7442
6	96800	4033.2	220.0	4840.0	11430	8066	7528
7	97875	4078.0	222.4	4893.8	11557	8156	7611
8	98951	4122.8	224.8	4947.6	11684	8245	7694
9	100026	4167.6	227.3	5001.3	11811	8335	7777
10	101102	4212.4	229.7	5055.0	11938	8424	7860
11	102177	4257.3	232.2	5108.7	12065	8514	7944
8. 0	103253	4302.2	234.6	5162.4	12192	8604	8028

N. B—No Allowance is made here for Pillars and Waste.

General ESTIMATES *on the* COMMON FIRE ENGINES.

(The Valve Engines require the Building about 3 feet 9 inches lower, and are not quite so expensive.)

	CYLINDER, 30 *In. Diameter.*				CYLINDER, 40 *In. Diameter.*				CYLINDER, 50 *In. Diameter*				CYLINDER, 60 *In. Diameter.*				CYLINDER, 70 *In. Diameter.*			
	Quantity.	L.	S.	D	Quantity.	L.	S.	D.	Quantity.	L.	S.	D.	Quantity.	L.	S.	D	Quantity.	L.	S.	D
FOUNDATION CUTTING,	—	3	—	—	—	4	—	—	—	5	—	—	—	6	—	—	—	7	—	—
Bricks,	55 ths.	44	—	—	70 ths.	56	—	—	90 ths.	72	—	—	120 ths.	96	—	—	150 ths.	120	—	—
Lime, (supposing the walls run with putty)	18 ch.	11	14	—	23 ch.	14	19	—	30 ch.	19	10	—	40 ch.	26	—	—	54 ch.	35	2	—
Sand,	36 lds.	3	12	—	46 lds.	4	12	—	60 lds.	6	—	—	80 lds.	8	—	—	108 lds.	10	16	—
Masons Bill,	—	22	—	—	—	26	—	—	—	39	—	—	—	55	—	—	—	75	—	—
Wood of Doors, Windows, Stairs laid in Walls, Floors and Roof,	—	33	—	—	—	38	—	—	—	44	—	—	—	50	—	—	—	56	—	—
Cylinder Beams,	16½ feet.	1	4	9	37 ft.	3	4	9	58 ft.	5	16	—	90 ft.	10	2	6	115 ft.	14	7	[illegible]
Regulator Beams,	80 ft.	9	6	8	110 ft.	15	2	6	140 ft.	22	15	—	175 ft.	35	—	—	210 ft.	52	10	—
Beam Heads,	27 ft.	2	5	0	35 ft.	3	4	2	44 ft.	4	19	—	54 ft.	6	15	—	65 ft.	8	18	9
Making and Sawing the Regulator Beams,	—	3	3	—	—	4	4	—	—	5	5	—	—	6	6	—	—	7	7	—
Spring Beams and pieces for packing the foundation of Regulator Beam,	58 ft.	4	2	2	74 ft.	5	4	10	112 ft.	7	18	8	132 ft.	9	7	—	154 ft.	10	18	2
Plugg Tree,	8 ft.	—	16	8	8 ft.	—	16	8	8 ft.	—	16	8	8 ft.	—	16	8	8 ft.	—	16	8
Cylinders,	26 cwt.	32	10	—	34 cwt.	42	10	—	50 cwt.	62	10	—	66 cwt.	82	10	—	76½ cwt.	95	12	6
Cylinder Bottoms,	8 cwt.	10	0	—	11 cwt.	13	15	—	17 cwt.	21	5	—	25 cwt.	31	5	—	30 cwt.	37	10	—
Pillars,	2, 1½ cw.	1	1	—	2, 1½ cw.	1	1	—	4, 4 cwt.	2	10	—	4, 4½ cw.	3	3	—	4, 5½ cw.	3	17	—
Injection Pipes,	8 cwt	5	12	—	9 cwt.	6	6	—	10 cwt.	7	—	—	11 cwt.	7	14	—	12 cwt.	8	8	—
Do. in the Cylinder,	½ cwt.	—	7	—	½ cwt.	—	7	—	½ cwt.	—	7	—	½ cwt.	—	7	—	½ cwt.	—	7	—
Snift Clack,	—	—	12	—	—	—	14	—	—	—	16	—	—	—	18	—	—	—	18	—
Pistons,	4 cwt.	2	16	—	8 cwt.	5	12	—	14 cwt.	9	16	—	22 cwt.	15	8	—	32½ cwt.	22	15	—
Piston Weights with Iron Handles,	1½ cwt.	1	1	—	2¼ cwt	1	11	6	2¾ cwt.	1	18	6	3½ cwt.	2	9	—	4½ cwt.	3	3	—
Gudgeon of Beam,	3½ cwt.	2	2	—	5½ cwt	3	6	—	7½ cwt.	4	10	—	11 cwt.	6	12	—	12¾ cwt.	7	13	—
Seat of Do.	2 cwt.	1	4	—	3 cwt.	1	16	—	4 cwt.	2	8	—	5 cwt.	3	—	—	6 cwt.	3	12	—
Receivers,	9¼ cwt.	6	9	6	10¾ cwt.	7	10	6	18¼ cwt.	12	15	6	22½ cwt.	15	15	—	27½ cwt	19	5	—
Brass work of Regulators, and fitting up,	56 lbs.	4	6	4	70 lbs.	5	7	11	84 lbs.	6	9	6	108 lbs.	8	6	6	136 lbs.	10	9	8
Communicating Pipes,	5 cwt.	3	10	—	8 cwt.	5	12	—	13 cwt.	9	2	—	18 cwt.	12	12	—	20 cwt	14	—	—
Rings to Do.	1¼ cwt.	—	17	6	1½ cwt.	1	1	—	2¾ cwt.	1	18	6	3 cwt.	2	2	—	3½ cwt.	2	9	—
Sink Pipe,	2½ cwt.	1	15	—	3 cwt.	2	2	—	3¼ cwt	2	5	6	3½ cwt.	2	9	—	3¾ cwt.	2	12	6
Lid to Do. and fitting up,	—	1	—	—	—	1	10	—	—	2	—	—	—	2	10	—	—	3	—	—
Hotwell,	7 cwt.	5	12	—	7½ cwt	6	—	—	7¾ cwt	6	4	—	8 cwt.	6	8	—	8½ cwt.	6	16	—
Feeding Pipes and Buoy Pipes,	2½ cwt.	1	15	—	3 cwt.	2	2	—	5 cwt.	3	10	—	6 cwt.	4	4	—	7 cwt.	4	18	—
Waste Water Do.	8 cwt.	5	12	—	8 cwt.	5	12	—	9 cwt.	6	6	—	9 cwt.	6	6	—	10 cwt.	7	—	—
Boilers,	1, 10½ d.	96	18	—	1, 14 di.	167	19	—	2, 12½ d.	271	15	—	2, 14½ d	361	16	6	2, 17 di.	496	14	—
Manhole,	5 cwt.	3	10	—	5 cwt.	3	10	—	10 cwt.	7	—	—	10 cwt.	7	-	—	10 cwt.	7	0	—
Grate Bars, and Bearing Bars,	23 cwt.	13	16	—	36 cwt.	21	12	—	59¼ cwt	35	11	—	74 cwt.	44	8	—	96 cwt	57	12	—
Plates before Fire, 2 below and 1 above,	6½ cwt.	3	18	—	9 cwt.	5	8	—	16 cwt.	9	12	—	19 cwt.	11	8	—	24 cwt.	14	8	—
Door Frames,	5½ cwt.	3	6	—	5½ cwt.	3	6	—	10¾ cwt.	6	9	—	10¾ cwt.	6	9	—	10¾ cwt.	6	9	—
Brass work (exclusive of Regulator)	—	8	—	—	—	12	10	—	—	13	—	—	—	16	10	—	—	19	10	—
Piston Shanks, main Chains, Martingals, &c.	—	41	5	—	—	67	3	7½	—	99	6	4½	—	137	3	6	—	175	2	10½
Plumber and Glazier's Bill,	—	30	—	—	—	40	—	—	—	53	—	—	—	80	—	—	—	90	—	—
Jack Head sett of Pumps compleat,	—	28	—	—	—	34	—	—	—	42	—	—	—	56	—	—	—	70	[illegible]	—
Cisterns. (wood only)	23 ft.	1	12	7	27 ft.	1	18	3	30 ft.	2	2	6	36 ft.	2	11	—	—	3	—	—
Outside Spring Frame and Top Pieces,	45½ ft.	3	4	5½	60 ft.	4	5	—	73 ft.	5	3	5	95 ft.	6	14	7	—	8	5	—
Paint, Oil, Paper, Flannel, Horn, spun Yarn, and Flannels for Men,	—	4	—	—	—	6	—	—	—	8	—	—	—	10	—	—	—	12	16	—
Sawing (exclusive of Regulator Beam,)	—	17	1	6½	—	21	8	8½	—	25	16	6½	—	31	1	—	—	37	14	10
Carpenter work of all, (Regulator Beam excepted, including models making,	—	40	14	—	—	52	—	—	—	65	—	—	—	84	—	—	—	108	7	—
Labourers, Extras, and Agency,	—	78	7	10	—	96	15	7	—	105	6	4	—	177	12	9	—	261	19	6½
		600	00	00		827	00	00		1146	00	00		1556	00	00		2022	00	00

	Thirty Inches CYLINDER.	*Forty Inches* CYLINDER.	*Fifty Inches* CYLINDER.	*Sixty Inches* CYLINDER.
	Depth 20 Fathoms, Lift 18 Fathoms, 11 Inches Bore, Pit $7\frac{1}{2}$ feet Diam.	Depth 30 Fathoms, Lift 26 Fathoms, 12 Inches Bore, Pit $7\frac{1}{2}$ feet Diam.	Depth 40 Fathoms, Lift 36 Fathoms, 13 Inches Bore, Pit $7\frac{1}{2}$ feet Diam.	Depth 50 Fathoms, Lift 47 Fathoms, 14 Inches Bore, Pit 8 feet Diam.
Sinking the Pits, 4l, 5l, 6l, & 7l. per fathom, with Engine Wright, - - - - - -	80£.	150£.	240£.	350£.
Timbering with deals 1 in. thick, Cribs and Nails,	40	60	80	100
Pumps compleat, with Bolts, &c. - - -	123	240	340	460
Spears and Y compleat, with 2 Buckets & 2 Clacks,	19	43	58	80
Buntons, Stays, with launder Boxes and fixing, -	18	35	48	70
Shear Legs compleat, with Malleable Iron, - -	3	5	8	10
Capston compleat of Wood, Cast Iron, Sheaves, &c.	25	39	50	55
And supposing the off-take drift to cost - -	20	32	34	60
Expence of the Shafts, - - -	328	604	858	1185
Expence of the Engines as above, -	600	827	1146	1556
Expence of so much of the Winning,	928	1431	2004	2741

OBSERVATIONS

ON THE

Foregoing Fire Engine and Colliery Estimates.

THESE Estimates of the expence of building Steam Engines, are applicable to those used in manufactories as well as collieries, and include every article in the Engine House, and take in the outside main chain of the regulator beam and spring frame also; and the prices are fixed agreeable to the general rate of charges in this neighbourhood, at the present time, supposing all the articles to be manufactured in a compleat and substantial manner, and the erections made near a Foundery. The bricks are laid at 16s. per thousand, lime 13s. per chaldron, timber from 1s. 8d. to 5s. per foot for the largest regulator beam, deal timber 17s. per foot, cylinders and cylinder bottoms 25s. per cwt. house water pipes 14s. per cwt. grate bars and beam gudgeons 12s. per cwt. Hotwells 16s. per cwt. and boilers 40s. per cwt.; and whoever wishes for information on this subject, will draw a comparison of the price of the above articles with that in his own neighbourhood.

With

With reſpect to the eſtimates on new winnings or openings of the ſundry depths above ſet forth, the local circumſtances are ſo various, that little uſeful information can be ſuppoſed to be derived from them; but as it frequently happens that gentlemen who undertake works of this kind, having little or no knowledge on the ſubject themſelves, wiſh to form ſome idea before the undertaking is commenced; if they will make allowance for all other expences neceſſary to compleat the winning, which every ſituation muſt point out for itſelf, ſuch as boring, coals to work the engine, enginemen, agency, corves, ropes, roads, buildings, machines, gins, ſtaiths, ponds, &c. &c. the eſtimates herewith annexed will be of ſome ſervice to them, ſuppoſing the ſize of the engine annexed to the ſeparate depths of the winnings are proper; but as to the real expence to be incurred in opening new collieries, it would be raſhneſs to attempt to ſet them forth with any tolerable degree of accuracy, when the place for the winning is even particularly pointed out, much leſs to attempt to give the expence in the way of general eſtimates.

There are doubtleſs ſundry caſes in this kingdom wherein the article of the engine pit expences, particularly ſpecified in the eſtimate, may be got executed for a much leſs ſum, and where even a leſs engine would be ſufficient to drain the work, and others may coſt even more than thoſe ſums; and if larger or additional engines be wanted, the expences of courſe muſt be greatly encreaſed.

FINIS

ERRATA.

Page.	Line.	
11,	2,	for *occular* read *ocular*.
13,	9,	for *ſtoneing* read *ſtoning*.
14,	10,	for *gates* read *gaits*.
24,	5,	for *thiner* read *thinner*.
45,	7,	for *that ſtands* read *it ſtands*.

T. WHEELHOUSE, TYPOGRAPHER.